Final Report of Project

"Plant Diversity Assessment and Establishment of Germplasm Bank for Conservation, Domestication and Breeding in *Capparis decidua*"

Under

Dr.B.P. Pal National Environment Fellowship Award for
Biodiversity for the year 2006

Submitted by

Dr. KULVIR SINGH BANGARWA
Professor
Department of Forestry
CCS Haryana Agricultural University
Hisar 125 004, India

Contents

Sr. No.	Topics	Pages
1	Front	1
2	Contents	3
3	(Part-I): "Plant Diversity Assessment and Establishment of Germplasm Bank for Conservation, Domestication and Breeding in *Capparis decidua*"	5
4	(PART-II): Preface	7-9
5	Abstract of the Project	11-14
6	Highlights of the Findings Achieved in the Project	15-16
7	Detailed Report of Work Done on the Project	17-81
	(A) Assessment of Diversity in Natural Population of *Capparis Decidua*	19
	(B) Documentation of traditional knowledge	47
	(C) Identification of promising reproductive material of *Capparis decidua* from different fragile environmental conditions.	56
	(D) Establishment of Germplasm Bank	64
	(E) Domestication and Breeding of *Capparis decidua*	79
8	Likely Impact of the Work on the Scientific Potential of Our Country	82-83
9	Bibliography	84-91
10	Executive Summary of the Project	92-99
11	(Part-III): Recommendation including remedial measures relevant to the environmental problems studied under the scheme	100

FINAL TECHNICAL REPORT OF R&D PROJECT DONE UNDER THE FELLOWSHIP AWARD SCHEME
(Part-I)

1 Title of the Project: "Plant Diversity Assessment and Establishment of Germplasm Bank for Conservation, Domestication and Breeding in *Capparis decidua*"

2 Name of the Awardee: Dr. K. S. BANGARWA
   & Address
       Professor
       Department of Forestry
       CCS Haryana Agricultural University
       Hisar 125 004, India

3 Number & date of sanction letter: vide No.16-1/2007-RE dated 11.11.2008

4 Duration of the Award:

(a) Date of Commencement: 10.12.2008

(b) Date of Completion: 9.12.2010

5 Budget:

Total amount sanctioned during the entire tenure under different sub-heads.

Sr no	Sub head	Receipt	Expenditure
1	Fellowship	Rs. 240000	Rs. 240000
2	Contingency	Rs. 170000	Rs. 125086
3	Salary of Research Associate	Rs. 335140	Rs. 329746
	Total	Rs. 745140	Rs. 694830
	Balance Amount	Rs. 50310	
	Bank interest etc.	Nil	
	Permanent equipment acquired	Nil	

(PART-II)

Preface

Biodiversity is the variety of genetic material, species and ecosystems found in nature. Collectively, biodiversity stabilizes our atmosphere and climate, protects water catchments and renews the soil. It also helps to keep ecosystems 'adaptable', should environmental conditions change abruptly. The diversity of nature is the foundation of the world's material wealth. From biodiversity, we develop food crops and derive the raw inputs and genetic materials for industry, agriculture and medicine. These benefits are worth many billions of dollars each year, and people spend further billions to appreciate nature and its diversity through tourism and recreation. Biodiversity also provides valuable indirect services through natural ecosystems. Traditional knowledge plays an important role in the conservation of biodiversity and its traditional uses. Traditional knowledge transforms biodiversity into bio-resources. Biodiversity and associated traditional knowledge are an integral strength of today's developing countries particularly in the areas of agriculture and Horticulture. It holds great potential all over the world that is increasingly being sensitized to traditional knowledge. Indigenous men and women over generations have bred races of several food, cash crops and Horticultural crops out of wild plants of the forests called landraces or local or indigenous varieties and these are the basic foundations of modern plant breeding and global food security. Indigenous farming communities have also identified and managed a series of genes through selection and cross breeding. These genes have potential traits of pest(s) and disease(s) resistance, drought tolerance, high salt tolerance, cold tolerance, tolerance to water logging etc. To develop a crop that can withstand global warming and climate changes across agricultural zones, international scientists visit tropical regions for crop varieties that are drought tolerant/ resistance and for this purpose they depend largely on traditional knowledge and local farmers.

Given the remarkable cultural diversity of India, the needs of human and livestock populations and substantial resources demands of the growing economy, conserving India's heritage of bio-diversity is an important task. Recognizing the importance of the subject of Biodiversity and with a view to further develop, deepen, and strengthen the expertise available in the country, the Advisory Committee under the Chairmanship of Dr. S.Z. Qasim, Member (Science), Planning Commission which met on 16th June, 1993 recommended institution of a National Fellowship in the area of Bio-diversity and deemed it appropriate to name it after Dr. B.P. Pal, firstly to bring a high standard to the fellowship and also to commemorate the extra-ordinary contributions of Dr. B.P. Pal in the subject. Dr B P Pal (1906-1989) a great scientist, institution builder is venerated for laying the foundations of India's Green Revolution in post independence times through focused translation of knowledge driven research into field performance by implementation of sound agriculture strategies. Dr B P Pal was the first Indian Scientist to become the Director of Indian Agriculture Research Institute, and later the first Director General of Indian Council of Agriculture Research. It was a matter of pride for me for receiving Dr BP Pal National Environment Fellowship Award for Biodiversity for the year 2006 to work on **"Plant Diversity Assessment and Establishment of Germplasm Bank for Conservation, Domestication and Breeding in *Capparis decidua*"**. I am thankful to Ministry of Environment & Forest for awarding me the most prestigious fellowship. The board of management of the University in its 225th meeting held on 7.8.2008 granted me permission to avail the fellowship on part time basis. I am grateful to Worthy Vice Chancellor and Board of Management of the University for providing me the opportunity to work on project under Dr BP Pal National Environment Fellowship Award for Biodiversity. I am thankful to the Director of Research of the University, Comptroller of the University and Professor & Head, Department of the Forestry of the University for providing me the facilities to work on **"Plant Diversity Assessment and Establishment of Germplasm Bank for Conservation, Domestication and Breeding in *Capparis decidua*"** under Dr BP Pal National Environment Fellowship

Award for Biodiversity for the year 2006. *Capparis decidua* has proven to be an economically important plant in southern Haryana, Rajasthan and elsewhere. It provides varied food and medicinal uses, building materials, fuel wood, and other income-generating opportunities. It contributes to environmental sustainability due to its soil-binding capacity and its ability to improve the soil fertility of sand dunes and to reduce soil alkalinity. Extensive research and support activities are thus needed to maximize the production, propagation, and utilization of this species to help contribute to rural livelihood and enhancement of desert lands.

(Dr Kulvir Singh Bangarwa)
Professor
Department of Forestry
CCS Haryana Agricultural University
Hisar 125004

Abstract of the Project

Capparis decidua is a shrub and an important medicinal plant. It is naturally found in varied habitats and has very good soil binding capacity. It improves the fertility of sand dunes and reduces alkalinity very sharply. Pickle and cooked vegetable of its unripe fruits form an integral part of human diet for stomach troubles in tropical and sub-tropical regions. Unripe green fruits are sold as hot cake in different regions. Top shoots and young leaves of the plant are used as plaster for boils and swelling, in powder form to relieve tooth troubles including pyorrhea. Fruits and seeds are used in cholera, dysentery and urinary purulent discharges. Root parts are useful in boils, eruptions, swelling, chronic and foul ulcers, cough and asthma. Wood is hard, heavy, and resistant to termites.

The project work on "Plant Diversity Assessment and Establishment of Germplasm Bank for Conservation, Domestication and Breeding in *Capparis decidua*" under Dr B.P.Pal National Environment Fellowship Award for Biodiversity was started in December 2008 for two years for the assessment of plant diversity, rate of loss in plant diversity and causes for decline in biodiversity of *Capparis decidua, d*ocumentation of traditional knowledge, folk varieties and other notable material, identification of promising reproductive material of *Capparis decidua* from different fragile environmental conditions, establishment of germplasm bank of *Capparis decidua* and domestication and breeding of *Capparis decidua*.

A survey of southern Haryana and Rajasthan was conducted during flowering and fruiting seasons. *Capparis decidua* has its existence in undisturbed areas. *Salvadora oleoides* has been very closely associated tree species of *Capparis decidua*. *Prosopis cineraria*, *Azadirachta indica* and *Calotropis procera* were also observed to be associated species. *Prosopis juliflora* (which regenerates faster and grows aggressively) has become a serious threat for the existence of our indigenous plants/trees and *Capparis decidua* is a serious victim of *Prosopis juliflora*. The flower buds and unripe green fruits of *Capparis decidua* are pickled and also cooked and eaten as vegetable. Hence, the local people used to go for over exploitation of *Capparis decidua*. Poor or lack of seed production due to over exploitation continues to be major cause for declining population of *Capparis decidua*. Flowering and fruiting in *Capparis decidua* have been observed twice in a year. Lot of variation was observed for flowering and fruiting between regions and between plants with in region. The flowers of *Capparis decidua* were observed to be complete having all the four parts viz., calyx (four sepals), corolla (four petals), androecium (many stamens usually 13-16 with a length varies from 0.5 to 1.2 cm) and gynoecium (ovary superior, bicarpellary syncarpous, unilocular with many ovules arranged on parental placenta, style absent and stigma about 1.1 cm.). In *Capparis decidua* stigma is closely surrounded by anthers. Pollination may be before flower opening or after flower opening. But the position of anthers in relation to stigma seems to ensure self-pollination. The seeds of *Capparis decidua* showed lot of variation for seed germination (fresh seeds) ranging from zero percent to 63.89%. The observations clearly suggested the extent of variation for seed dormancy among different seed lots of *Capparis decidua*. Traditional Uses index of *Capparis decidua* was worked out in comparison to other trees/shrubs of arid region. Index score analysis on the basis of simultaneous consideration of medicinal value, food value and potential for sand dune stabilization and salt tolerance suggested the superiority of *Capparis decidua* in comparison to *Azadirachta indica*, *Prosopis cineraria* and *Acacia nilotica*.

The flower buds and unripe green fruits of *Capparis decidua* are pickled and also cooked and eaten as vegetable. Pickle and cooked vegetables of unripe fruits are very useful for stomach troubles especially for constipation. The *Capparis decidua* has been traditionally useful in the treatment of coughs, asthma, bronchial inflammation, respiratory disorders, indigestion, pyorrhea, intermittent fevers, rheumatism, ulcers, boils, vomiting, piles, bone fracture inflammation and acute pain, scorpion bite, antidote to poison, toothache, diabetes, heart, liver and kidney problems, trachoma (Chronic conjunctivitis that can cause blindness), infected hooves, scabies and eczema. It is being used as laxative, tonic and mouth freshner. Being non shrinkage nature of wood, it is specifically used as central liver (kila and mani) in household chaki. *Capparis decidua* is of much use in climate prediction and features in farmers' strategies in natural resources production and management and agricultural planning. The farmers consider that if blooming in *Capparis decidua* is greater and flowers are deep pink, then the temperature is more than 45°C and the rainfall will be less than normal. Based on the observations of the numbers of flowers and fruits and the canopy of *Capparis decidua*, the farmers select their crop varieties and cropping systems for the following rainy season. Large sized trees up to 8.5 metres with higher yield of fruit per plant were observed in Salasar, Sikar (Rajasthan), Kalwas, Hisar (Haryana) and Balsamand, Hisar (Haryana). Plants mostly shrubs with large spread and higher yield of fruit per plant were obsereved in Sahewala, Hisar (Haryana), Bapoda, Bhiwani (Haryana), Fatehpur, Sikar (Rajasthan), Salasar, Sikar (Rajasthan), Ghodaria Khurad, Sikar (Rajasthan), Nokhda, Bikaner (Rajasthan), Phagalwas, Sikar (Rajasthan) and Dadar, Churu (Rajasthan). Seedlings of nine progenies were transplanted in the field following Randomized Block Design (RBD) with three replications during September, 2009. The plant height varies from 31.2 cm to 43.5 cm with a mean of 37.5 cm whereas plant spread varies from 27.5 cm to 39.4 cm with a mean of 35.8 cm at the age of fifteen months after transplanting. The root length was observed higher than plant height and plant spread. The root length varies from 39.4 cm to 86.5 cm with a mean of 65 cm. Root length varies from 1.26 times of plant height to 2.04 times of plant height with a mean of 1.73 times of plant height whereas root length varies from 1.43 times of plant spread to 2.06 times of plant spread with a mean of 1.81 times of plant spread. Coefficient of variation for root length was 25.7 percent whereas for plant height and plant spread, coefficients of variations were 11.06 per cent and 15.7 per cent, respectively. Seedlings of thirty-nine accessions collected from southern Haryana, adjoining Rajasthan and Bikaner region of Rajasthan were transplanted in the field following RBD during February, 2010. The highest plant height of 24 cm was observed in progeny no 51 from Kalwas, Hisar (Haryana) followed by progeny no 17 and 16 from Nokhda, Bikaner (Rajasthan), progeny no 40 from Bapoda, Bhiwani (Haryana), progeny no 28 from Sahewala, Hisar (Haryana) and progeny no 4 from Hawala, Rajsamand (Rajasthan) nine months after transplanting. The progeny no 51 from Kalwas, Hisar (Haryana) and progeny no 28 from Sahewala, Hisar (Haryana) were observed comparatively more erect growing. Seedlings of seventy-seven accessions collected from southern Haryana, adjoining Rajasthan and Bikaner region of Rajasthan were transplanted in the field following Randomized Block Design (RBD) with three replications during September, 2010. The highest plant height of 20.6 cm was observed in accession no 142 from Salasar, Sikar (Rajasthan) followed by accession no 51 from Kalwas, Hisar (Haryana), accession no 40 from Bapoda, Bhiwani (Haryana), accession no 16 from Nokhda, Bikaner (Rajasthan), accession no 118 from Nokhda, Bikaner (Rajasthan) and accession no 128 from Fatehpur, Sikar (Rajasthan) three months after transplanting. The plant height at the age of three months after transplanting varies from 11.4 cm in accession no 33 from Loharu, Bhiwani (Haryana) to 20.6 cm in accession no 142 from Salasar, Sikar (Rajasthan) with a mean of 15.52 cm and 13.55 % coefficient of variation.

Capparis decidua can be successfully grown from seeds and suckers. Its matured fruits which have turned pink should be collected either in May-June or August-October. Seeds should be separated out by washing the pulp of fruits with in 2-3 days. This will enhance the viability of seeds. Seeds are to be sown in polythene bags (with 3:1 ratio of sandy soil and FYM) in August if seeds are collected in May-June or February if seeds are collected in August-October. Collection of seeds in August-October and sowing of seeds in early February are desirable. Transplanting of seedlings should be done in early February by choosing seedlings of more than 15 cm height. Suckers of *Capparis decidua* are to be collected in early February and should be established in polythene bags (with 3:1 ratio of sandy soil and FYM). Plants raised from suckers attain usually attain a height of more than 15 cm in 4-5 months and hence these plants are transplanted in field during rainy season.

Highlights of the Findings Achieved in the Project

- *Capparis decidua* has its existence in undisturbed areas. *Salvadora oleoides* has been very closely associated tree species of *Capparis decidua*. *Prosopis juliflora* is a serious threat for the existence of *Capparis deciduas*. The local people used to go for over exploitation of *Capparis decidua* to collect fruits from inner depth of shrub even by damaging the branches. Poor or lack of seed production continues to be major cause for declining population of *Capparis decidua*.

- Lot of variation was observed for flowering and fruiting between regions and between plants with in region. The flowers of *Capparis decidua* were observed to be complete having all the four parts viz., calyx, corolla, androecium and gynoecium. In *Capparis decidua* stigma is closely surrounded by anthers. But the position of anthers in relation to stigma seems to ensure self-pollination. *Capparis decidua* is predominantly self-pollinating species.

- The observations for seed germination (zero percent to 63.89%) clearly suggested the extent of variation for seed dormancy among different seed lots of *Capparis decidua*.

- Index score analysis on the basis of simultaneous consideration of medicinal value, food value and potential for sand dune stabilization and salt tolerance suggested the superiority of *Capparis decidua* in comparison to *Azadirachta indica*, *Prosopis cineraria* and *Acacia nilotica*.

- The *Capparis decidua* has been traditionally useful in the treatment of coughs, asthma, bronchial inflammation, respiratory disorders, indigestion, pyorrhea, intermittent fevers, rheumatism, ulcers, boils, vomiting, piles, bone fracture inflammation and acute pain, scorpion bite, antidote to poison, toothache, diabetes, heart, liver and kidney problems, trachoma, infected hooves, scabies and eczema. It is being used as laxative, tonic and mouth freshner.

- Being non shrinkage nature of wood, wood of *Capparis decidua* is specifically used as central liver (kila and mani) in household chaki.

- *Capparis decidua* is of much use in climate prediction and features in farmers' strategies in natural resources production and management and agricultural planning. The farmers consider that if blooming in *Capparis decidua* is greater and flowers are deep pink, then the temperature is more than 45°C and the rainfall will be less than normal. Based on the observations of the numbers of flowers and fruits and the canopy of *Capparis decidua,* the farmers select their crop varieties and cropping systems for the following rainy season.

- Large sized trees up to 8.5 metres with higher yield of fruit per plant were observed in Salasar, Sikar (Rajasthan), Kalwas, Hisar (Haryana) and Balsamand, Hisar (Haryana). Plants mostly shrubs with large spread and higher yield of fruit per plant were obsereved in Sahewala, Hisar (Haryana), Bapoda, Bhiwani (Haryana), Fatehpur, Sikar (Rajasthan), Salasar, Sikar (Rajasthan), Ghodaria Khurad, Sikar (Rajasthan), Nokhda, Bikaner (Rajasthan), Phagalwas, Sikar (Rajasthan) and Dadar, Churu (Rajasthan).

- Progeny testing results of *Capparis decidua* suggested lot of variation for above ground and below ground growth and the root length was observed higher than plant height and plant spread. Erectness in plant growth was observed to be under genetic control.

- Early growth in *Capparis decidua* germplasm bank having thirty-nine and seventy-seven accessions showed lot of variation for growth pattern.

- It has been established that *Capparis decidua* can be successfully grown from seeds and suckers. Its matured fruits which have turned pink should be collected either in May-June or August-October. Seeds should be separated out by washing the pulp of fruits with in 2-3 days. Seeds are to be sown in polythene bags (with 3:1 ratio of sandy soil and FYM) in August if seeds are collected in May-June or February if seeds are collected in August-October. Collection of seeds in August-October and sowing of seeds in early February are desirable. Transplanting of seedlings should be done in early February by choosing seedlings of more than 15 cm height. Suckers of *Capparis decidua* are to be collected in early February and should be established in polythene bags (with 3:1 ratio of sandy soil and FYM). Plants raised from suckers attain usually attain a height of more than 15 cm in 4-5 months and hence these plants are transplanted in field during rainy season.

Detailed Report of Work Done on the Project

a) **Summary of the objectives**

i. Assessment of plant diversity, rate of loss in plant diversity and causes for decline in biodiversity of

Capparis decidua.

ii. Documentation of traditional knowledge, folk varieties and other notable material.

iii. Identification of promising reproductive material of *Capparis decidua* from

iv. different fragile environmental conditions.

v. Establishment of germplasm bank of *Capparis decidua*.

vi. Domestication and breeding of *Capparis decidua*.

b) **Methodology:**

i. Assessment of plant diversity of *Capparis decidua*

A survey of southern Haryana and Rajasthan was conducted during flowering and fruiting seasons. The data was recorded from natural population of *Capparis decidua* for the assessment of plant diversity. The rate of loss in plant diversity and causes for decline in biodiversity of *Capparis decidua* were worked out by collecting information from experienced local people of different places.

ii. Documentation of traditional knowledge

A survey of southern Haryana and Rajasthan was conducted and data with respect to usefulness of *Capparis decidua* were collected by involving local people of different regions.

iii. Identification of promising reproductive material of *Capparis decidua* from different environmental conditions

A survey of southern Haryana and Rajasthan was conducted and phenotypically superior plants of *Capparis decidua* growing under different conditions were selected and seeds were collected for establishment of progeny testing.

iv. Establishment of germplasm bank of *Capparis decidua*

A survey of southern Haryana and Rajasthan was conducted during flowering and fruiting seasons. Matured fruits of *Capparis decidua* were collected from fifty-five plants/shrubs/trees growing naturally in southern Haryana, adjoining Rajasthan and Bikaner District of Rajasthan during May 2009. Again, during November 2009, matured fruits were collected from 61 plants/shrubs/trees growing naturally in southern Haryana and adjoining Rajasthan. Seedlings of all the accessions were raised in nursery. Seedlings of seventy-seven and thirty-nine accessions collected from southern Haryana, adjoining Rajasthan and Bikaner region of Rajasthan were transplanted in the field following Randomized Block Design (RBD) with three replications.

v. **Domestication and breeding of *Capparis decidua***

Reproductive biology of *Capparis decidua* with respect to seed production age, flower development, breeding system, seed quality and seed storability were worked out. Seed collection method, nursery raising technique and planting technique were standardized based on different experiments/observations.

Results, Discussion and Analysis along with Tables/Charts and Figures

(A)　Assessment of Diversity in Natural Population of *Capparis Decidua*

(a)　Associated species of *Capparis decidua*: A survey of southern Haryana and Rajasthan was conducted during flowering and fruiting seasons. *Capparis decidua* has its existence in undisturbed areas. *Salvadora oleoides* has been very closely associated tree species (Plate 1) of *Capparis decidua*. *Prosopis cineraria*, *Azadirachta indica* and *Calotropis procera* were also observed to be associated species. Now a days, Prosopis *juliflora* an exotic tree species having very fast rate of growth is spreading like weeds particularly in undisturbed lands. In this way, undisturbed lands are common habitat for *Capparis decidua* and Prosopis *juliflora* (Plate 2).

Prosopis juliflora (which regenerates faster and grows aggressively) has become a serious threat for the existence of our indigenous plants/trees and *Capparis decidua* is a serious victim of *Prosopis juliflora*. The flower buds and unripe green fruits of *Capparis decidua* are pickled and also cooked and eaten as vegetable. The immature fruits are in great demand and have high economic value. Hence, the immature fruits are harvested and sold at high prices (upto Rs sixty per Kg). The local people used to go for over exploitation of *Capparis decidua*. They usually try to collect fruits from inner depth of shrub even by damaging the branches. The availability of matured seed is almost nil. This practice puts seed production and propagation of *Capparis decidua* at risk. Poor or lack of seed production continues to be major cause for declining population of *Capparis decidua*. Therefore, regeneration in *Capparis decidua* is through suckers only and that too is suppressed by *Prosopis juliflora*. Hence, *Prosopis juliflora* is seriously threatening the survival of *Capparis decidua* (Plate 3) and other indigenous species (Singh and Singh, 2011). If proper selection is made from available variability, *C. decidua* can make an excellent crop for extreme arid zone of Rajasthan and Gujarat states where a few species can survive and can be domesticated. *Capparis decidua* was found to be one of the best species for shelter belts to check the movement of sand in the Thar Desert of India (Pandey and Rokad 1992). As a drought resistant, it has relatively good nutritive value and withstands neglect. It is found in varied habitats and has very good. soil binding capacity which can be propagated and cultivated on large scale for checking wind erosion on sandy wastelands (Gupta *et al.*, 1989).

Plate 1a: *Capparis decidua and Salvadora oleoides* **very closely associated**

Plate 1b: *Capparis decidua and Salvadora oleoides* very closely associated

Plate 2a: Prosopis *juliflora* a threat for *Capparis decidua*

Plate 2: Prosopis *juliflora* **a threat for** *Capparis decidua*

Plate 3: Shrub of *Capparis decidua* suppressed by *Prosopis juliflora*

(b) **Flowering and fruiting in *Capparis decidua*:** Flowering and fruiting in *Capparis decidua* has been observed twice in a year. Flowering was observed during March-April and by the end of April fruit setting was completed. But in some parts, flowering was observed up to end of May. Flowering and fruiting were also observed in June in very limited no of plants. Regular flowering was also observed during August-September and by the end of September fruit setting was completed. Flowering and fruiting were also observed in October-November in very limited no of plants. Lot of variation was observed for flowering and fruiting between regions and between plants with in region. It was observed that some of the plants/shrubs/trees of *Capparis decidua* have one time flowering and fruiting in a year whereas others have two times flowering and fruiting in a year. Very few plants have alternate year flowering and fruiting. From a group of *Capparis decidua* plants/shrubs/trees at a particular place, some plants were observed to have full flowering whereas other plants were observed to have flowering ranging from zero to 80 %. Variation has been observed for flower colour (Plate 4). *Kair* fruits show variability in shape as well as size", broadly, the fruits are round or elongated, with a different diameter. Variability in flower colour in natural stands of C. *decidua* was reported with brick red- and yellow-coloured flowers (Singh and Singh, 2011). In the present scenario, where there is no commercial cultivation of this species, and green fruits are harvested by the rural people, it is obvious that more fruits will be picked up from the non-spiny open type shrubs that are earlier fruiting.

(c) **Type of flowers and mode of pollination in *Capparis decidua***

Mode of pollination in a plant species can be determined following the three steps.

The first step is to critically examine the flowers. Mechanisms like dioecy, monoecy, protogyny, protandry and cleistogamy are easily detected; they clearly indicate the mode of pollination. The flowers of *Capparis decidua* were observed to be complete having all the four parts (Plate 5) viz., calyx (four sepals), corolla (four petals), androecium (many stamens usually 13-16 with a length varies from 0.5 to 1.2 cm) and gynoecium (ovary superior, bicarpellary syncarpous, unilocular with many ovules arranged on parental placenta, style absent and stigma about 1.1 cm.).

Plate 4: Variation for flowering pattern and colour in *Capparis decidua*

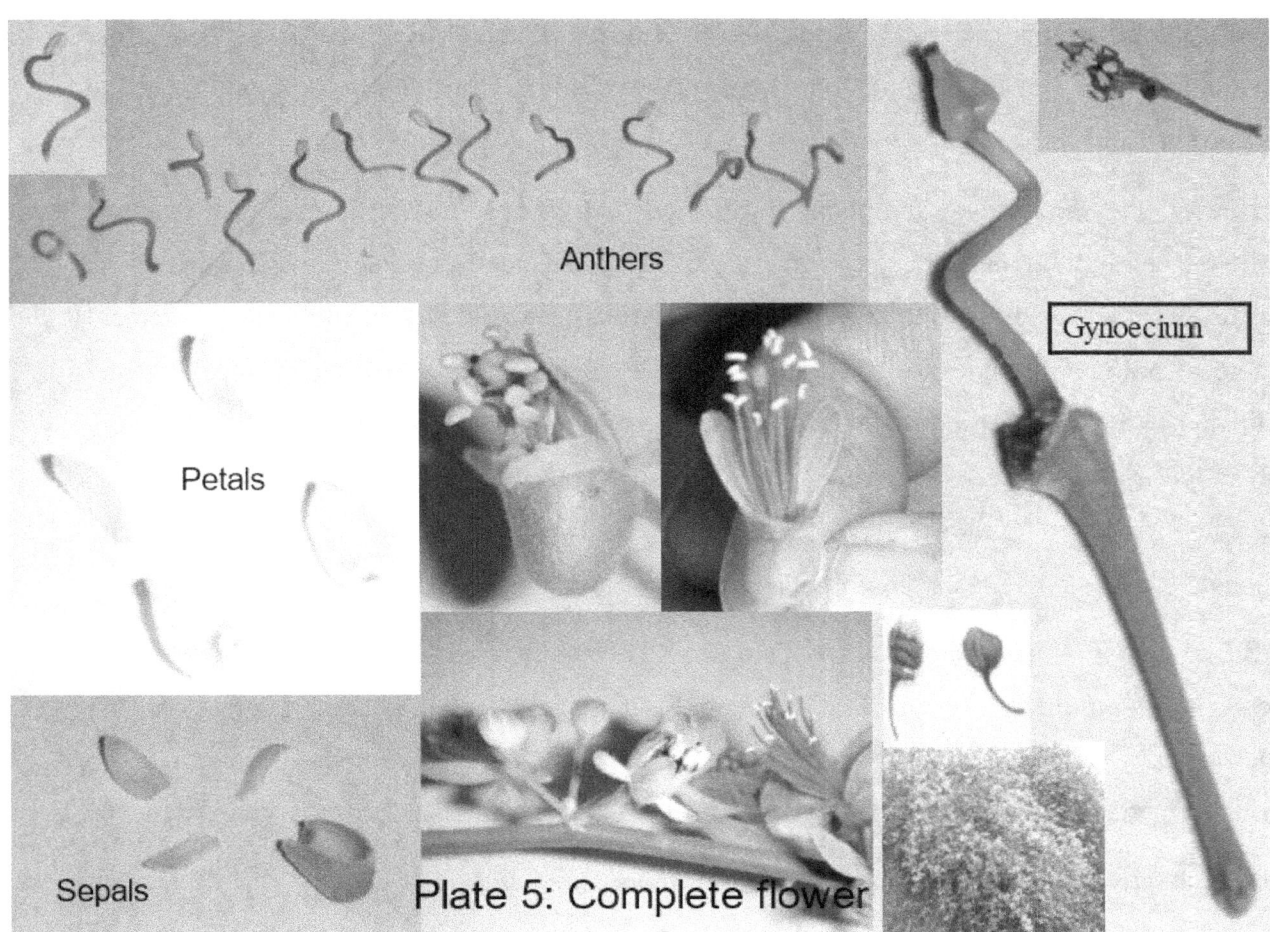

Plate 5: Complete flower

In *Capparis decidua* stigma is closely surrounded by anthers (Plate 6). Pollination may be before flower opening or after flower opening. But the position of anthers in relation to stigma seems to ensure self-pollination.

The second step is to isolate the plants/shrubs/trees individually and record seed set in them. Space isolation i.e., individual a plant growing at sufficient distance to prevent cross-pollination is preferable to isolation by bags or cages. The latter two situations may create an environment unfavourable to pollination and seed set. Failure to set seed in isolation proves the species to be cross-pollinated; however, setting of seeds is the only and sure indicator of self-pollination. Sufficient no of fruits and seeds were observed in a *Capparis decidua* plants/shrubs/trees in an isolation of about one km. Seeds collected from isolated plants showed normal germination and growth. Hence, *Capparis decidua is* predominantly self-pollinating species.

Finally, effects of selfing (inbreeding) on the vigor should be studied. Loss in vigor due to inbreeding is common in cross-pollinating species; self-pollinating species show no inbreeding depression. The seeds collected from isolated plants/shrubs/trees of *Capparis decidua* showed normal growth. These observations further supported the self-pollinated nature of *Capparis decidua.*

(d)   Seed collection from natural population of *Capparis decidua*

Matured fruits (plate 7) of *Capparis decidua* were collected from fifty-five plants/shrubs/trees growing naturally in southern Haryana, adjoining Rajasthan and Bikaner District of Rajasthan during May 2009. Again, during November 2009, matured fruits were collected from 61 plants/shrubs/trees growing naturally in southern Haryana and adjoining Rajasthan.

(e)   Variation for Seed Germination and Seed Dormancy in *Capparis decidua*

Collection of matured seed is a serious problem in *Capparis decidua* because its unripe fruits are in great demand for making pickle at commercially large scale. Fruits of *Capparis decidua* are also in great demand for medicinal uses. Poor landless people used to collect almost all the unripe fruits as these plants/shrubs/trees are naturally growing in undisturbed common lands. The fruits when turned pink were collected for the purpose of seeds collection (Plate 7). The fruits were washed and seeds were separated from the pulp. The seeds of *Capparis decidua* showed lot of variation for seed germination (fresh seeds) ranging from zero percent to 63.89%. About fifty percent seed lots showed increase in germination percent up to three

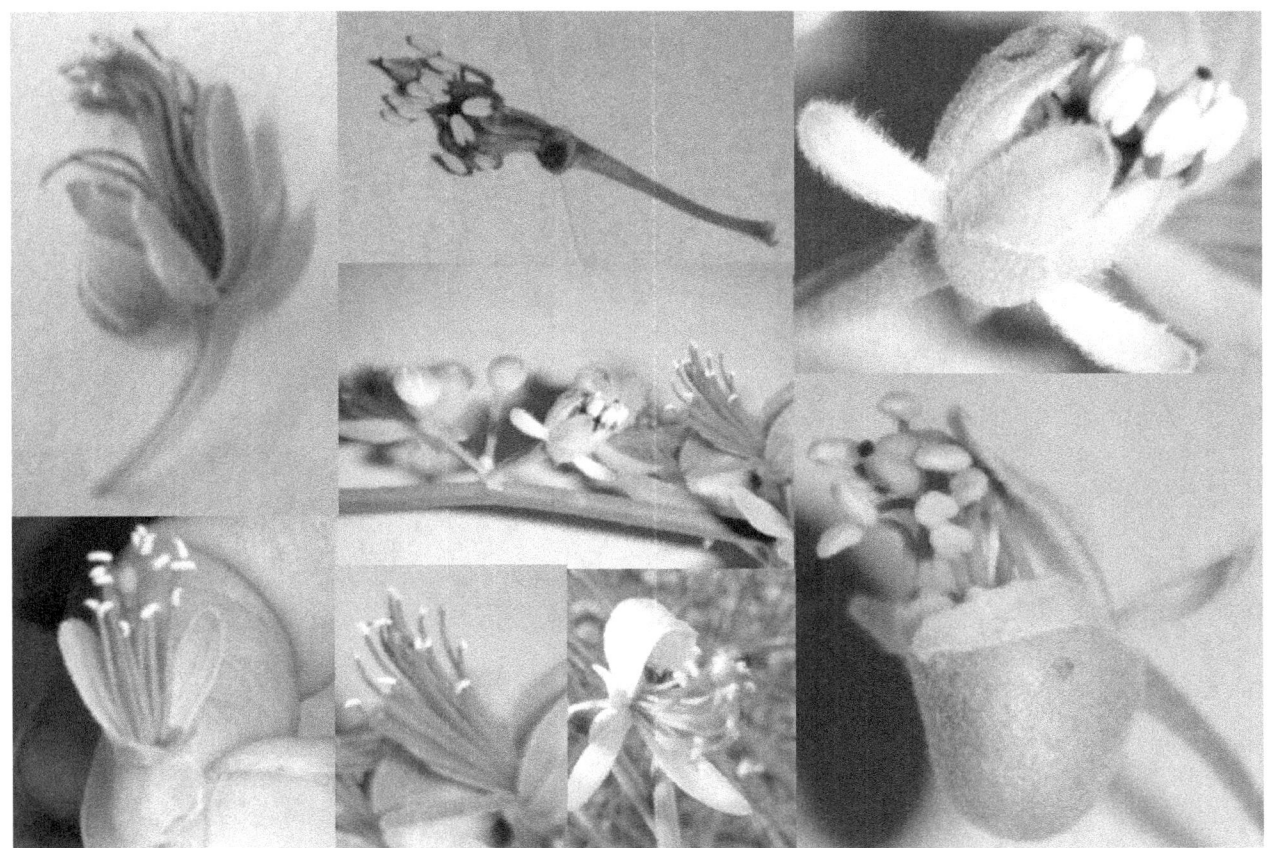

Plate 6: Stigma surrounded by anthers in *Capparis decidua*

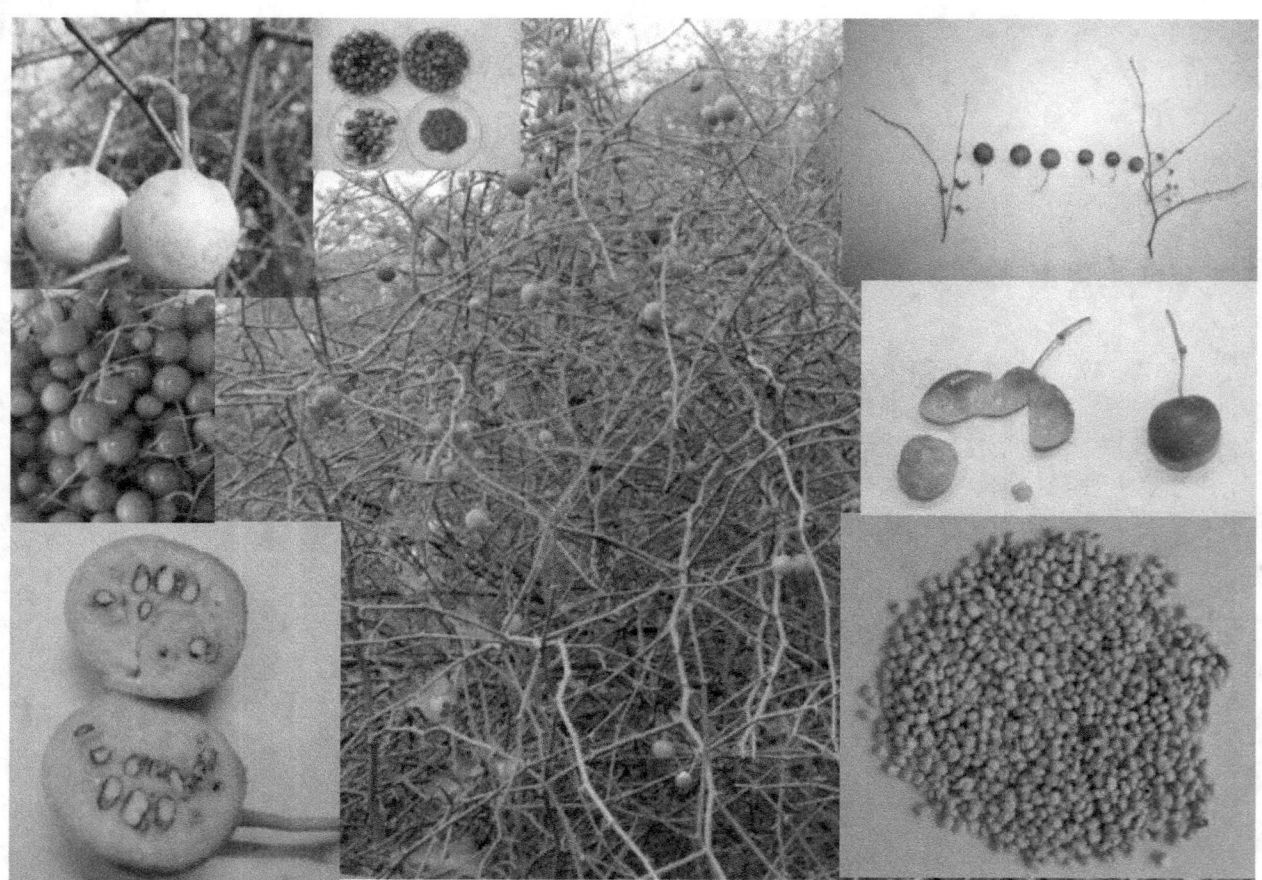

Plate 7: Matured fruits & seeds in *Capparis decidua*

months after storage. This is clear indication of presence of dormancy in different seed lots of *Capparis decidua*. Four seed lots from diverse regions showed increase in germination percent up to six months consistently (Fig. 1). These observations clearly suggested the extent of variation for seed dormancy among different seed lots of *Capparis decidua*.

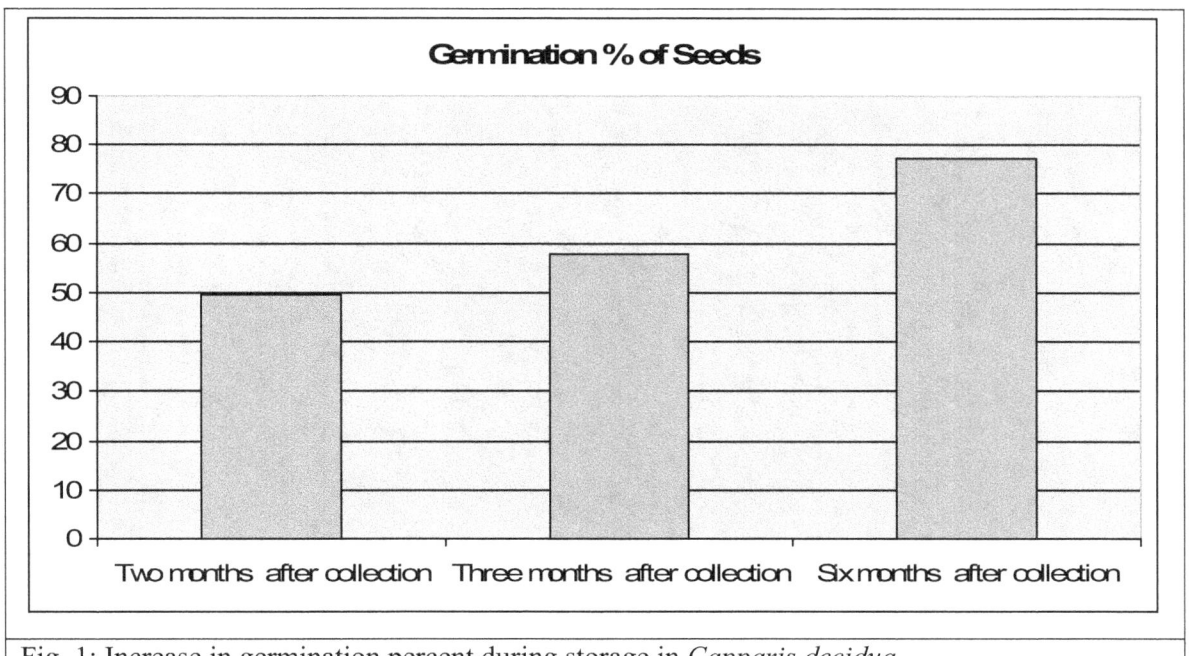

Fig. 1: Increase in germination percent during storage in *Capparis decidua*

(f) Assessment of diversity in natural population of *Capparis decidua*

The mean, range and CV for morphological and seed characters in *Capparis decidua* are presented in Table 1 and diversity is described character wise.

(i) Tree height (m): Tree height varies (Fig 2) from 1.5 to 8.5 m in Salasar, Sikar with a mean of 2.75 m. Majority of *Capparis decidua* plants were observed to be bushy in nature. But lot of variation (cv = 34.6%) was observed for plant height. The growth pattern in natural population of *Capparis decidua* has been shown in plate 8. Under natural populations, rich genetic diversity with a wide range of variability occurs in *Capparis decidua in* habit, fruit size, colour of fruits, petals, pulp content, spiny habit, spreading of branches and compactness of canopy, and time of flowering and fruiting. In general, two distinct plant types of kair occur: a tree form, which is relatively unusual, and a shrub form, in the majority of plants (Singh and Singh, 2011). It appears that tree form is attained when the plant grows from seed and remains undisturbed. On the other hand, plants exposed to biotic interference may tend to produce more shoots and also propagate through root suckers. This view is supported by the fact that *C. decidua* occurred singly in tree form and mostly in clusters in bush form. The plant can also be trained by allowing a single stem to grow, and can be integrated as one of the components in different cropping models. However, scientific evaluation is needed to assess the relative compatibility and economic viability under different resource situations.

Table 1 Mean, Range and CV for Morphological and Seed Characters in *Capparis decidua*

Acc. No.	Location	Tree height (m)	Tree diameter (cm)	Crown Diameter (m)	No of Fruiting trees/10 trees	Fruit diameter (cm)	Seed diameter (mm)	100 seed weight	No of seeds per fruit	Range of seeds per fruit
1	Khatana Khera, Ajmer (Rajasthan)	2.0	0.0	0.73	9	-	3.39	2.47	7.5	1-14
2	Khatana Khera, Ajmer (Rajasthan)	2.5	0.0	1.02	9	-	3.20	1.85	12.5	2-23
3	Hawala, Rajsamand (Rajasthan)	3.0	0.0	1.11	8	13.87	3.38	2.19	13.5	4-23
4	Hawala, Rajsamand (Rajasthan)	2.6	0.0	0.89	8	15.06	3.33	1.82	11.5	1-20
5	Managaliawas, Ajmer (Rajasthan)	2.7	0.0	1.08	2	-	3.51	1.64	10.0	3-17
6	Hawala, Rajsamand (Rajasthan)	1.8	0.0	1.24	8	-	3.21	1.34	15.0	8-22
7	Gagal, Ajmer (Rajasthan)	1.9	0.0	1.31	4	12.54	3.77	2.12	10.0	2-18
8	Palukalam, Jaipur (Rajasthan)	2.1	0.0	1.15	8	15.33	3.78	2.12	12.0	6-18
9	Hawala, Rajsamand (Rajasthan)	2.4	0.0	0.92	8	-	3.58	1.57	13.5	8-19
10	Managaliawas, Ajmer (Rajasthan)	1.7	0.0	0.76	2	12.78	3.46	1.40	10.0	5-15
11	Palukalam, Jaipur (Rajasthan)	2.1	0.0	1.08	4	13.92	4.02	2.70	14.5	3-26
12	Gajner, Bikaner (Rajasthan)	3.0	0.0	2.86	4	15.23	3.51	2.12	10.0	1-19
13	Kolayat, Bikaner (Rajasthan)	3.0	0.0	3.31	4	17.51	3.47	2.16	11.0	5-17
14	Raneri, Bikaner (Rajasthan)	3.0	9.5	2.55	4	15.47	3.41	2.13	9.5	4-15
15	Raneri, Bikaner (Rajasthan)	3.3	0.0	0.78	5	16.21	3.43	2.16	11.0	5-17
16	Nokhda, Bikaner (Rajasthan)	2.7	0.0	0.86	4	16.76	3.40	2.04	9.5	3-16

17	Nokhda, Bikaner (Rajasthan)	3.3	0.0	3.98	4	17.02	3.37	2.08	11.0	2-20
18	Nokhda, Bikaner (Rajasthan)	2.4	0.0	1.56	4	18.08	3.39	2.21	11.0	2-20
19	Bhana, Bikaner (Rajasthan)	3.6	0.0	3.50	4	17.22	3.69	2.33	12.5	3-22
20	Raneri, Bikaner (Rajasthan)	3.0	0.0	3.18	5	16.43	3.68	2.01	17.5	1-34
21	Raneri, Bikaner (Rajasthan)	2.7	0.0	3.18	5	17.40	3.50	1.95	10.5	4-17
22	Raneri, Bikaner (Rajasthan)	2.1	0.0	1.78	4	16.46	3.71	2.28	8.0	2-14
23	Raneri, Bikaner (Rajasthan)	3.9	23.9	1.91	4	14.19	3.38	2.12	12.5	2-23
24	Raneri, Bikaner (Rajasthan)	3.6	20.1	3.50	4	15.03	3.57	2.22	10.5	3-18
25	Muklan, Hisar (Haryana)	3.6	0.0	4.14	2	15.78	3.82	2.22	9.0	3-15
26	Bheria, Hisar (Haryana)	2.4	0.0	3.18	2	12.59	3.68	2.15	10.5	3-18
27	Chaudhariwas, Hisar (Haryana)	1.9	0.0	2.13	5	16.24	3.60	2.22	9.0	2-16
28	Sahewala, Hisar (Haryana)	4.2	35.0	3.21	4	16.56	3.62	2.05	10.0	1-19
29	Sahewala, Hisar (Haryana)	2.3	0.0	1.85	4	16.92	3.39	1.78	10.0	3-17
30	Baddumughal, Bhiwani (Haryana)	2.4	15.9	2.86	3	16.18	3.64	1.76	8.5	3-14
31	Kharkarhi, Bhiwani (Haryana)	3.0	0.0	4.66	4	16.75	3.74	2.50	11.0	3-19
32	Dhani Tohae, Bhiwani (Haryana)	3.0	0.0	3.18	5	17.14	3.60	2.17	20.5	6-35
33	Loharu, Bhiwani (Haryana)	2.7	0.0	6.05	4	14.49	3.48	1.98	12.0	2-22
34	Madhogarh, Mohindergarh (Haryana)	2.1	0.0	3.18	2	15.97	3.51	2.24	10.0	2-18
35	Dhanoda, Mohindergarh (Haryana)	3.0	0.0	4.14	3	14.35	3.22	1.76	9.5	2-17
36	Dhanoda, Mohindergarh (Haryana)	2.4	0.0	4.55	3	16.66	3.99	2.80	8.5	3-14

37	Saharanwas, Rewari (Haryana)	2.7	0.0	4.14	3	15.60	3.51	2.32	10.5	4-17
38	Mirch, Bhiwani (Haryana)	3.3	0.0	4.55	2	12.07	3.68	2.28	9.5	3-16
39	Bapoda, Bhiwani (Haryana)	1.8	0.0	1.34	5	17.58	3.87	2.56	10.5	2-19
40	Bapoda, Bhiwani (Haryana)	3.6	50.9	2.67	5	18.00	3.77	2.50	9.5	1-18
41	Bapoda, Bhiwani (Haryana)	1.8	0.0	4.14	5	17.32	3.59	1.85	11.5	2-19
42	Premnagar, Bhiwani (Haryana)	1.8	0.0	2.71	3	12.75	3.22	2.38	11.5	2-21
43	Jatu Luhari, Bhiwani (Haryana)	2.1	0.0	2.13	6	15.75	3.44	1.85	11.0	3-19
44	Jatu Luhari, Bhiwani (Haryana)	2.1	0.0	3.92	6	18.32	3.35	1.78	12.5	2-23
45	Kalwas, Hisar (Haryana)	3.0	28.6	3.09	4	16.17	3.60	2.54	9.5	1-18
46	Rawat khera, Hisar (Haryana)	3.0	38.2	3.50	4	16.68	3.17	1.63	11.0	3-19
47	Kalwas, Hisar (Haryana)	3.6	35.0	5.54	4	14.22	3.25	1.80	12.5	6-19
48	Kalwas, Hisar (Haryana)	3.6	57.3	4.46	4	18.07	3.60	2.10	9.0	3-15
49	Chirod, Hisar (Haryana)	2.1	19.1	3.06	3	19.46	3.51	2.35	12.0	2-22
50	Chirod, Hisar (Haryana)	2.7	22.3	4.14	3	19.56	3.27	1.68	13.0	3-23
51	Kalwas, Hisar (Haryana)	6.2	47.7	2.77	2	20.76	3.43	2.06	20.5	4-37
52	Salem garh, Hisar (Haryana)	3.0	0.0	4.30	4	17.84	4.11	2.85	18.0	4-32
53	Salem garh, Hisar (Haryana)	1.6	0.0	2.58	4	14.17	3.70	1.93	19.0	2-36
54	Sunda was, Hisar (Haryana)	1.8	36.6	3.60	3	18.67	3.77	1.77	18.0	4-32
55	Sunda was, Hisar (Haryana)	1.9	0.0	1.08	3	14.79	2.85	2.77	9.5	1-18
101	Arya nagar, Hisar (Haryana)	2.1	0.0	1.21	3	-	3.35	1.53	8.5	2-15
102	Arya nagar,	2.2	0.0	1.56	3	-	3.30	1.67	10.0	3-17

	Hisar (Haryana)								
103	Hisra city (Haryana)	1.8	0.0	1.62	2	-	3.57	1.82	7.0	2-12
104	Khedar, Hisra (Haryana)	1.7	0.0	1.21	3	-	3.20	1.73	10.5	3-18
105	Rawat khera, Hisar (Haryana)	2.1	0.0	1.18	6	13.68	3.51	1.75	10.5	3-18
106	Rawat khera, Hisar (Haryana)	2.4	0.0	2.86	6	16.20	3.36	1.95	11.0	3-19
107	Rawat khera, Hisar (Haryana)	2.7	0.0	3.18	6	15.79	3.64	2.13	11.0	3-19
108	Rawat khera, Hisar (Haryana)	2.4	0.0	2.23	6	16.66	3.88	2.02	10.0	3-17
109	Rawat khera, Hisar (Haryana)	3.0	0.0	1.59	6	16.14	3.44	1.95	10.5	3-18
110	Rawat khera, Hisar (Haryana)	3.0	0.0	2.23	6	12.39	3.36	1.63	10.0	3-17
111	Kalwas, Hisar (Haryana)	3.6	0.0	3.98	4	15.91	3.50	1.85	11.5	5-18
112	Kalwas, Hisar (Haryana)	2.7	0.0	4.30	4	17.96	3.47	1.82	12.0	5-19
113	Kalwas, Hisar (Haryana)	4.2	0.0	3.18	4	15.67	3.39	2.04	12.0	5-19
114	Kalwas, Hisar (Haryana)	2.1	0.0	1.50	4	18.45	3.53	1.92	11.5	5-18
115	Talwandi Ruka, Hisar (Haryana)	3.3	0.0	3.02	3	16.14	3.72	2.02	11.0	5-17
116	Talwandi Ruka, Hisar (Haryana)	3.6	0.0	2.86	3	16.43	3.50	2.24	11.0	5-17
117	Bure, Hisar (Haryana)	3.0	0.0	2.07	2	14.27	3.62	2.04	10.0	4-16
118	Dubeta, Hisar (Haryana)	4.5	0.0	3.82	2	17.42	3.84	1.80	10.0	5-15
119	Rajgarh (Rajasthan)	2.4	0.0	4.46	2	18.08	3.45	1.50	9.0	3-15
120	Dadar, Churu (Rajasthan)	3.9	47.7	4.93	7	16.64	3.63	2.01	10.5	4-17
121	Dadar, Churu (Rajasthan)	2.1	0.0	1.97	7	18.81	3.87	2.60	11.5	3-20
122	Dadar, Churu (Rajasthan)	2.7	0.0	3.02	7	19.34	3.70	2.72	10.5	3-18
123	Dadar, Churu (Rajasthan)	3.0	0.0	4.14	7	18.65	3.87	2.57	10.0	4-16
124	Dadar, Churu (Rajasthan)	1.7	0.0	2.61	7	18.87	4.01	2.90	10.5	3-18
125	Hochar, Sikar (Rajasthan)	2.4	0.0	1.91	3	17.17	3.83	2.28	9.5	3-16

126	Hochar, Sikar (Rajasthan)	2.1	0.0	4.14	3	19.18	3.87	2.32	9.0	3-15
127	Fatehpur, Sikar (Rajasthan)	3.9	0.0	3.18	5	19.34	3.82	2.38	10.5	4-17
128	Fatehpur, Sikar (Rajasthan)	4.2	0.0	4.77	5	17.24	3.54	2.20	11.0	4-18
129	Fatehpur, Sikar (Rajasthan)	4.5	54.1	3.09	5	17.48	3.49	1.52	10.0	3-17
130	Fatehpur, Sikar (Rajasthan)	2.1	0.0	2.20	5	20.08	3.60	2.00	10.5	4-17
131	Fatehpur, Sikar (Rajasthan)	3.0	31.8	2.01	2	18.58	3.73	2.00	11.5	4-19
132	Ghashu Madhopura, Sikar (Rajasthan)	2.7	0.0	4.77	3	19.16	3.97	2.45	12.0	4-20
133	Ghashu Madhopura, Sikar (Rajasthan)	2.2	0.0	2.39	3	18.30	4.16	2.45	11.0	5-17
134	Khudi, Sikar (Rajasthan)	1.5	0.0	2.04	2	16.70	3.76	2.04	10.5	4-17
135	Khudi, Sikar (Rajasthan)	3.9	22.3	2.86	2	17.76	3.64	2.27	10.5	3-18
136	Phagalwas, Sikar (Rajasthan)	2.4	0.0	3.18	7	18.56	3.75	2.10	11.0	4-18
137	Phagalwas, Sikar (Rajasthan)	3.0	0.0	2.55	7	19.21	3.49	1.90	11.0	4-18
138	Phagalwas, Sikar (Rajasthan)	2.7	0.0	4.77	7	19.25	3.84	2.26	10.0	3-17
139	Phagalwas, Sikar (Rajasthan)	2.7	0.0	6.37	7	20.03	3.67	2.42	11.0	5-17
140	Phagalwas, Sikar (Rajasthan)	2.4	0.0	2.99	7	19.35	3.87	2.55	10.0	4-16
141	Salasar, Sikar (Rajasthan)	4.5	0.0	3.18	2	18.94	3.81	2.98	14.0	6-22
142	Salasar, Sikar (Rajasthan)	8.5	66.8	3.02	5	19.94	3.78	2.64	18.0	7-29
143	Salasar, Sikar (Rajasthan)	3.0	0.0	6.68	5	20.20	3.88	2.62	13.5	4-23
144	Salasar, Sikar (Rajasthan)	2.7	0.0	4.46	5	20.68	3.85	2.94	15.1	6-25
145	Salasar, Sikar (Rajasthan)	2.4	0.0	2.23	5	20.61	3.49	1.95	12.0	3-21

146	Singdoh, Sikar (Rajasthan)	2.0	0.0	2.55	2	18.41	3.90	2.38	14.0	5-23
147	Ghodaria Khurad, Sikar (Rajasthan)	2.4	0.0	3.18	7	22.65	3.92	2.94	11.5	4-19
148	Ghodaria Khurad, Sikar (Rajasthan)	2.0	0.0	2.48	7	17.87	3.97	2.70	11.5	5-18
149	Ghodaria Khurad, Sikar (Rajasthan)	3.6	38.2	4.77	7	20.43	3.92	2.40	16.0	6-26
150	Ghodaria Khurad, Sikar (Rajasthan)	3.6	0.0	2.86	7	20.61	4.10	3.15	13.0	4-22
151	Ghodaria Khurad, Sikar (Rajasthan)	1.9	0.0	2.42	7	20.55	3.67	2.32	15.5	3-28
152	Ghodaria Khurad, Sikar (Rajasthan)	3.0	38.2	2.55	7	18.17	3.87	2.30	14.0	5-23
153	Chirawa, Jhunjhanu (Rajasthan)	1.8	0.0	2.90	3	17.62	3.68	1.93	16.0	4-28
154	Badi-sahi, Jhunjhanu (Rajasthan)	2.4	0.0	2.64	2	19.61	3.95	2.32	15.5	7-24
155	Shailu, Jhunjhanu (Rajasthan)	2.4	0.0	3.82	2	19.91	3.64	1.62	16.5	6-27
156	Hanumangarh (Rajasthan)	1.8	0.0	2.80	3	17.45	3.61	2.20	16.0	4-28
157	Hanumangarh (Rajasthan)	1.9	0.0	2.61	2	16.07	3.46	2.00	17.5	8-27
158	Hanumangarh (Rajasthan)	2.0	0.0	2.39	4	16.63	3.54	1.74	16.0	7-25
159	Hanumangarh (Rajasthan)	2.0	0.0	2.42	3	16.91	3.59	2.20	13.0	4-22
160	Hanumangarh (Rajasthan)	2.1	0.0	2.71	1	16.49	3.65	2.34	18.0	8-28
161	Khedar, Hisra (Haryana)	2.1	0.0	3.15	2	15.89	3.44	1.92	13.5	5-22
	Mean	2.8	35.2	2.88	4.3	17.10	3.60	2.09	11.8	
	Range	1.5-8.5	0-66.84	.73-6.68	1--9	12.07-22.65	2.85-4.16	1.34-3.15	7.0-20.5	
	SD	1.0	15.0	1.27	1.9	3.83	0.23	0.37	2.7	
	CV	34.6	42.5	43.75	43.95	22.39	6.38	17.22	23.1	

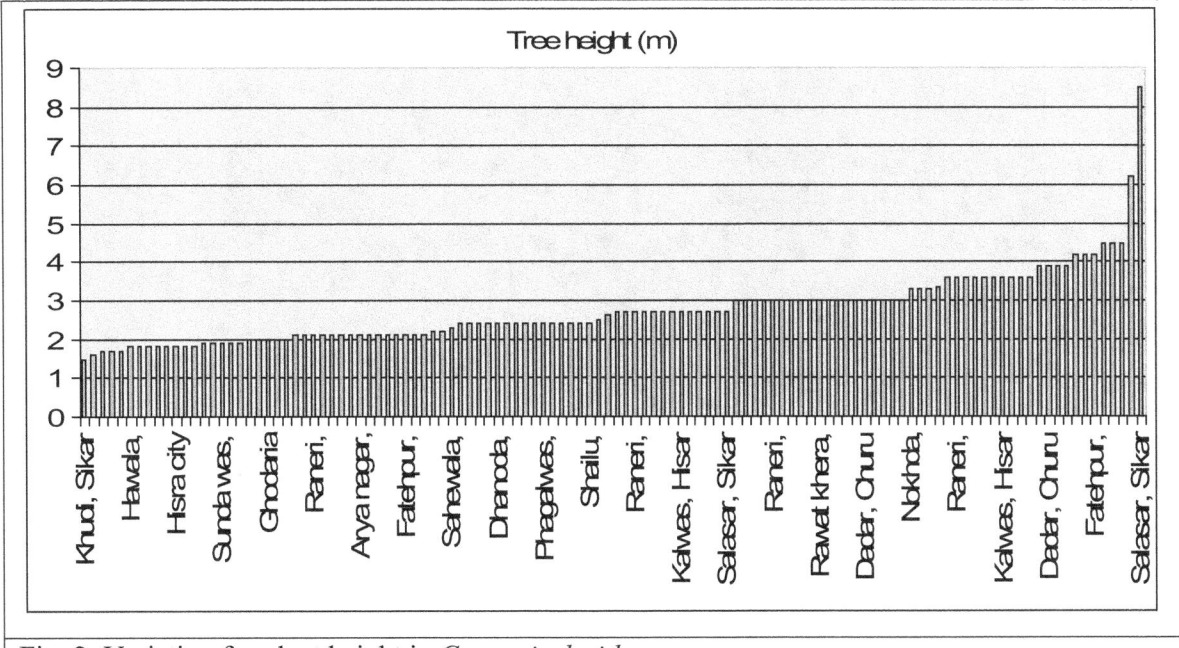

Fig. 2: Variation for plant height in *Capparis deciduas*

(ii) Tree diameter (cm): Majority of plants (96 out of 116) from where the fruits were collected and data were recorded, no well-defined stem was observed. The stem diameter was recorded from 20 plants/trees. The tree diameter varies (Fig 3) from 9.54 to 66.84 cm with a mean of 35.21cm. Sufficient amount of variation (cv = 42.51%) was observed for tree diameter. The growth pattern in natural population of *Capparis decidua* has been shown in plate 8.

(iii) Crown Diameter (m): Crown Diameter varies (Fig 4) from 0.73 in accession no 1 from Khatana Khera, Ajmer (Rajasthan) to 6.68 m in accession no 143 from Salasar, Sikar (Rajasthan) with a mean of 2.88 m. Majority of *Capparis decidua* plants/shrubs/trees were spreading type in nature. Lot of variation (cv = 43.75 %) was observed for plant height. The crown diameter was observed to be increased with age. The increase in diameter was also observed lengthwise. The growth pattern in natural population of *Capparis decidua* has been shown in plate 8.

Plate 8: Variation for height and growth pattern in *Capparis decidua*

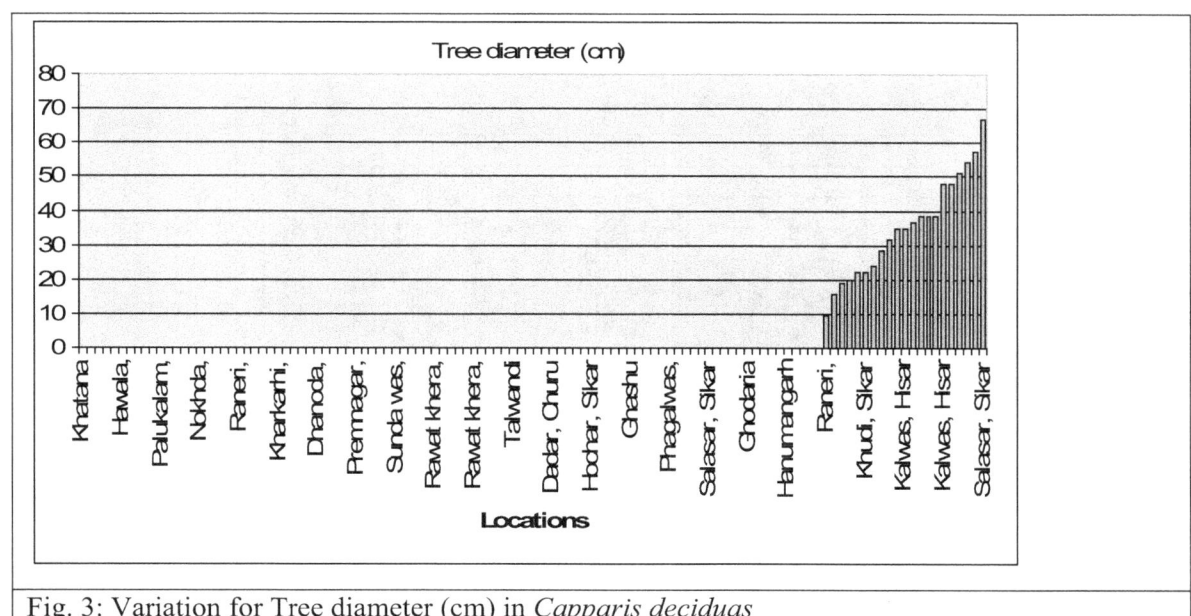

Fig. 3: Variation for Tree diameter (cm) in *Capparis deciduas*

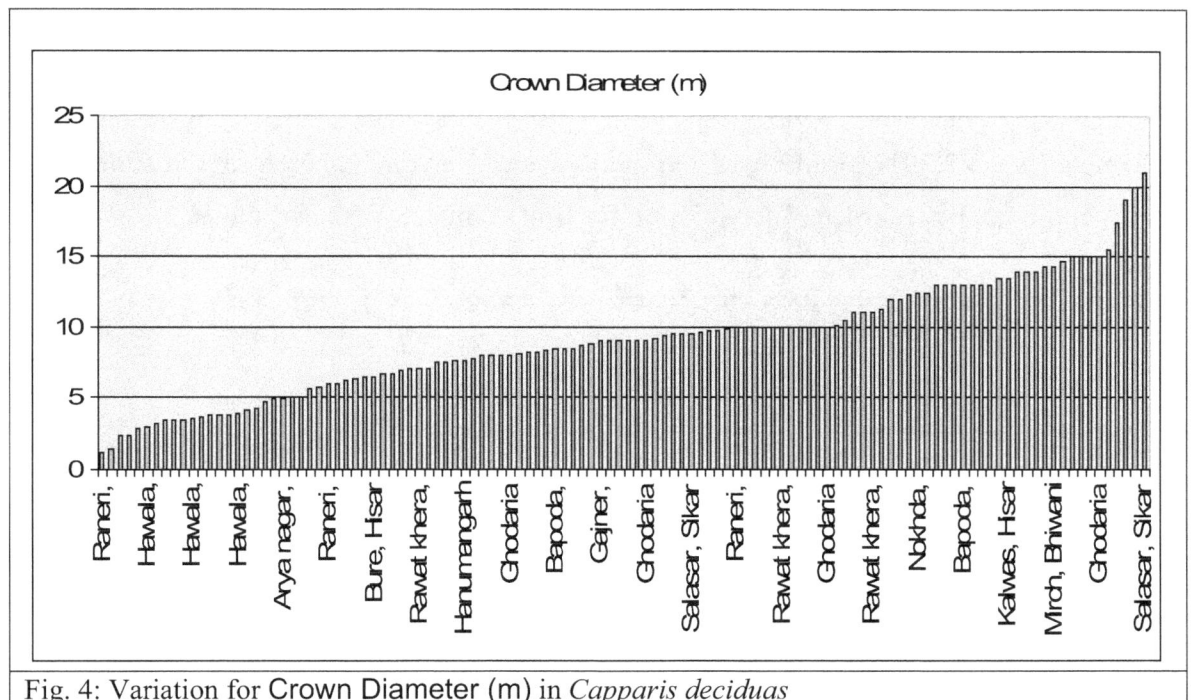

Fig. 4: Variation for Crown Diameter (m) in *Capparis deciduas*

(iv) No of Fruiting trees: No of Fruiting trees out of ten varies (Fig 5) from 1 to 9 with a mean of 4.3. Coefficient of variation was observed to be 43.95 percent for no of Fruiting trees. Highest variation was observed for flowering and fruiting in *Capparis decidua*. Moreover, from a group of *Capparis decidua* plants/shrubs/trees at a particular place, some plants were observed to have full flowering whereas other plants were observed to have flowering ranging from zero to 80 % (Plate 4).

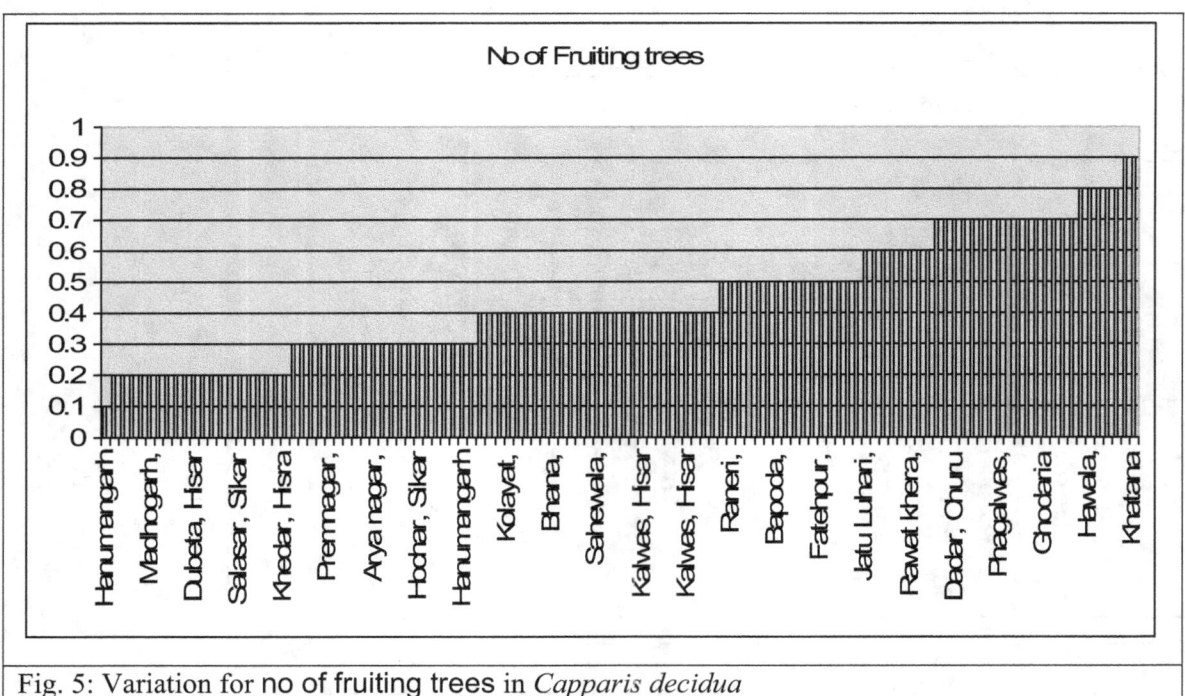

Fig. 5: Variation for no of fruiting trees in *Capparis decidua*

(v) Fruit diameter (cm): Fruit diameter varies (Fig 6) from 1.21-2.26 cm with a mean of 1.71 cm. Fruits of *Capparis decidua* plants/shrubs/trees were collected from 116 plants. The coefficient of variation for fruit diameter was 22.39%. Fruits from ten plants were over ripe. Therefore, the fruit diameter from ten trees could not be recorded. The variation for fruit diameter was continuous.

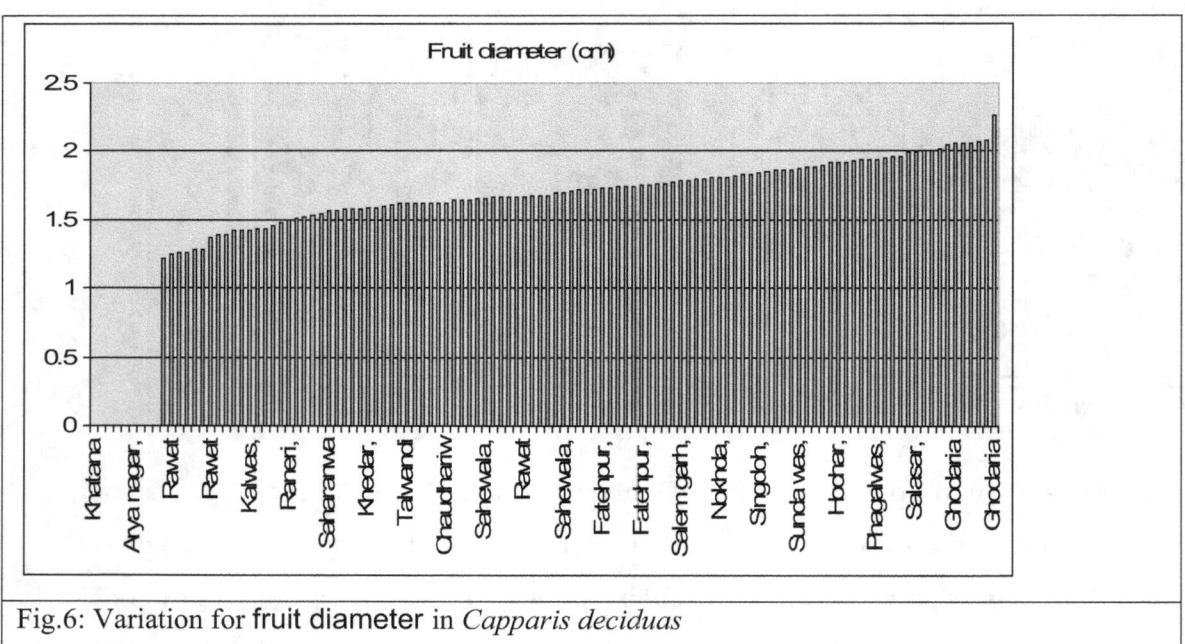

Fig.6: Variation for fruit diameter in *Capparis deciduas*

(vi) Seed diameter (mm): Seed diameter varies (Fig 7) from 2.85 to 4.16 mm with a mean of 3.6 mm. Seeds of *Capparis decidua* were collected from 116 plants/shrubs/trees. The coefficient of variation for seed diameter was 6.38%. Variation for seed diameter was comparatively lower and variation was continuous.

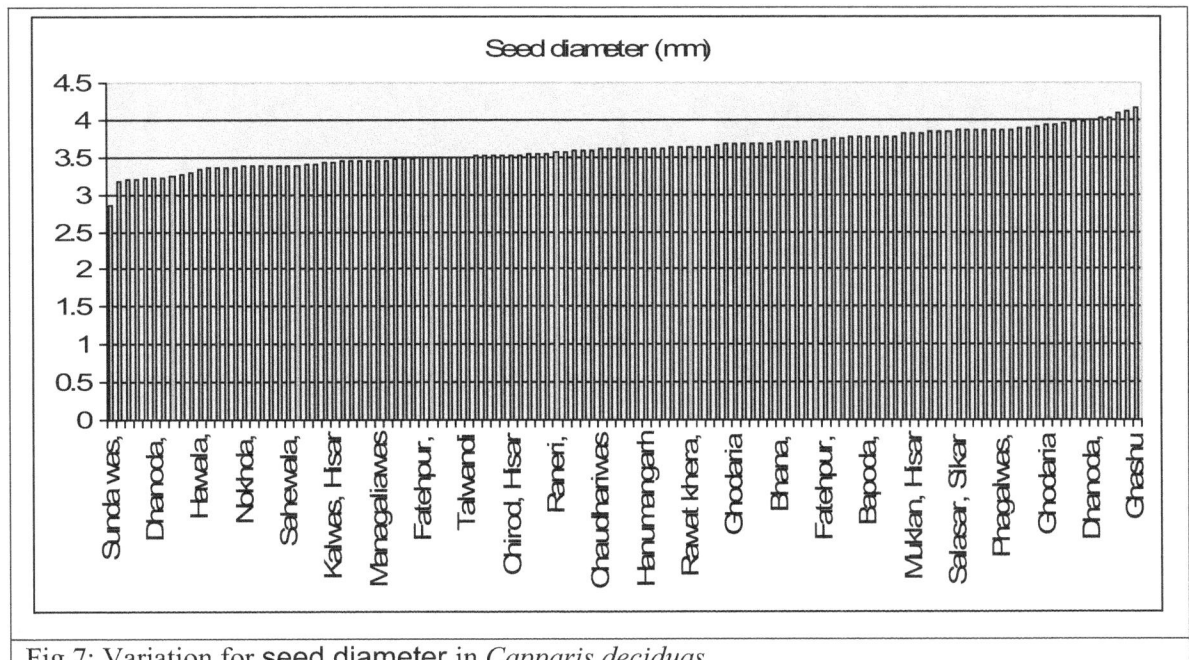

Fig.7: Variation for seed diameter in *Capparis deciduas*

(vii) 100 Seed weight (g): 100 Seed weight varies (Fig 8) from 1.34 to 3.15 g with a mean of 2.07 g. The highest value of 100 seed weight was 2.35 times of lowest value. The coefficient of variation for seed diameter was 17.22%. Variation for 100 Seed weight was also continuous.

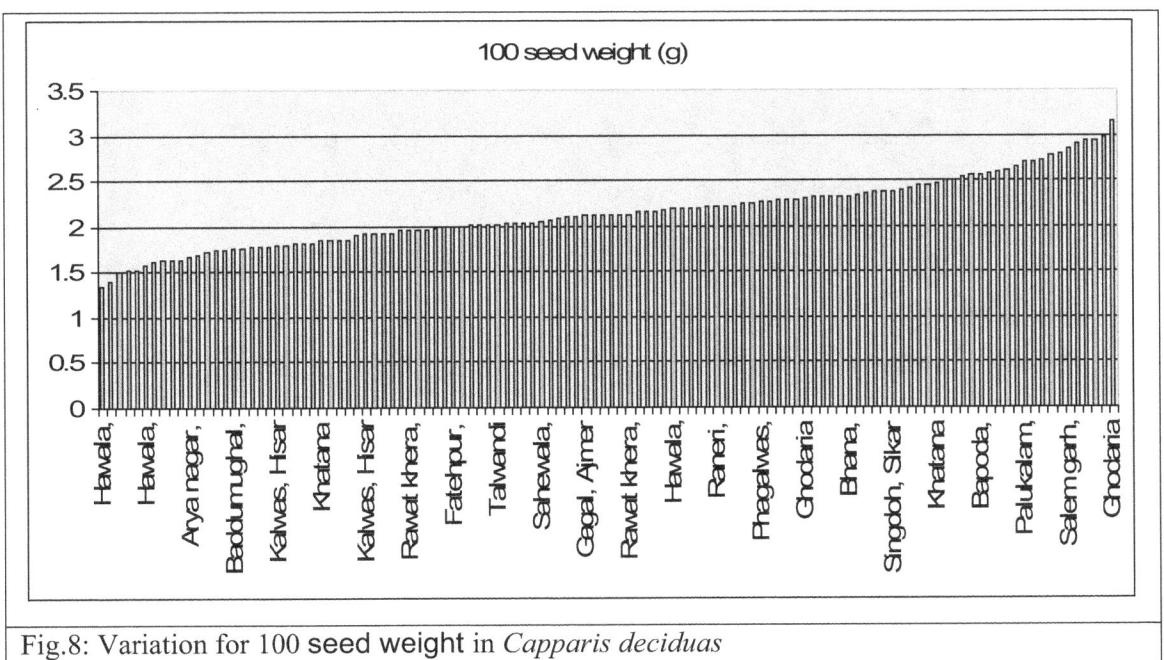

Fig.8: Variation for 100 seed weight in *Capparis deciduas*

(viii) No of seeds per fruit: No of seeds per fruit (on the basis of average) varies (Fig 9) from 7 to 20.5 with a mean of 11.81. The highest value of no of seeds per fruit was almost three times of lowest value. The coefficient of variation for no of seeds per fruit was 23.11 %. Variation for no of seeds per fruit was also continuous. But if no of seeds per fruit on the basis of individual fruits are considered then no of seeds per fruit varies from 1.0 to 37.0 with a mean of 11.81.

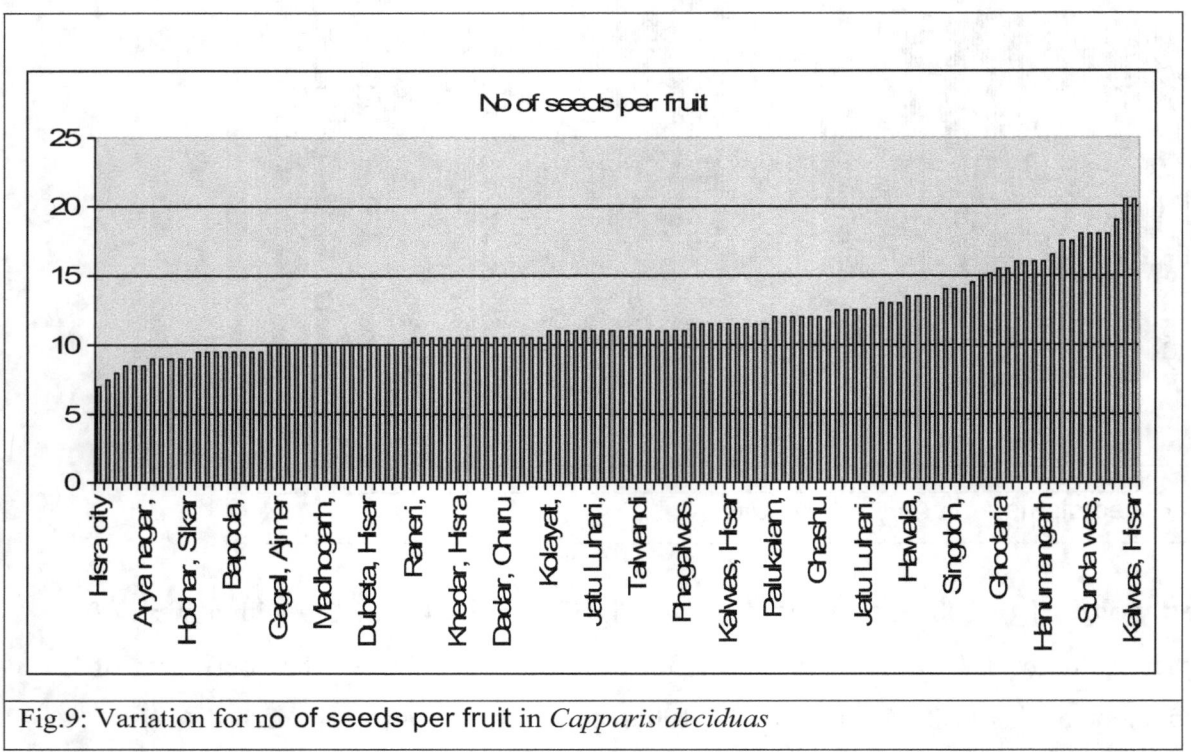

Fig.9: Variation for no of seeds per fruit in *Capparis deciduas*

(ix) Minimum no of seeds per fruit: Minimum no of seeds per fruit varies (Fig 10) from 1 to 8 with a mean of 3.65. The highest value of minimum no of seeds per fruit was eight times of lowest value. The coefficient of variation for minimum no of seeds per fruit was 44.11 percent. Maximum occurrences of minimum no of seeds per fruit was 3 followed by 4, 2, and 5.

(x) Maximum no of seeds per fruit: Maximum no of seeds per fruit varies (Fig 11) from 12 to 37 with a mean of 19.94. The highest value of maximum no of seeds per fruit was more than three times of lowest value. The coefficient of variation for maximum no of seeds per fruit was 24.42 %. Most frequently occurring maximum no of seeds per fruit were 17, 18 and 19.

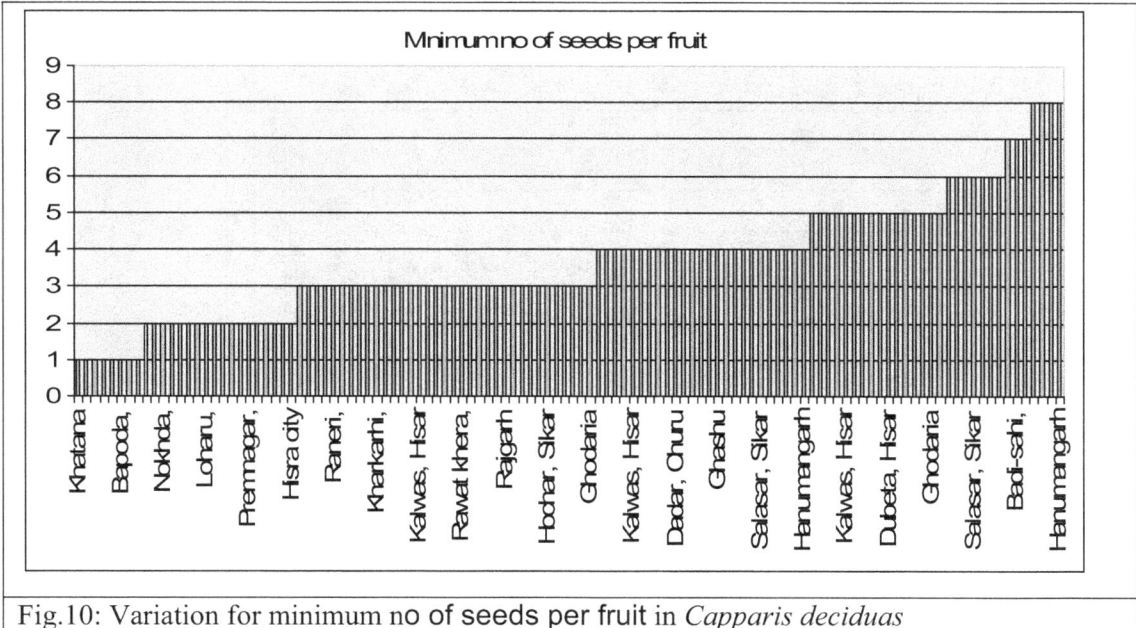

Fig.10: Variation for minimum no of seeds per fruit in *Capparis deciduas*

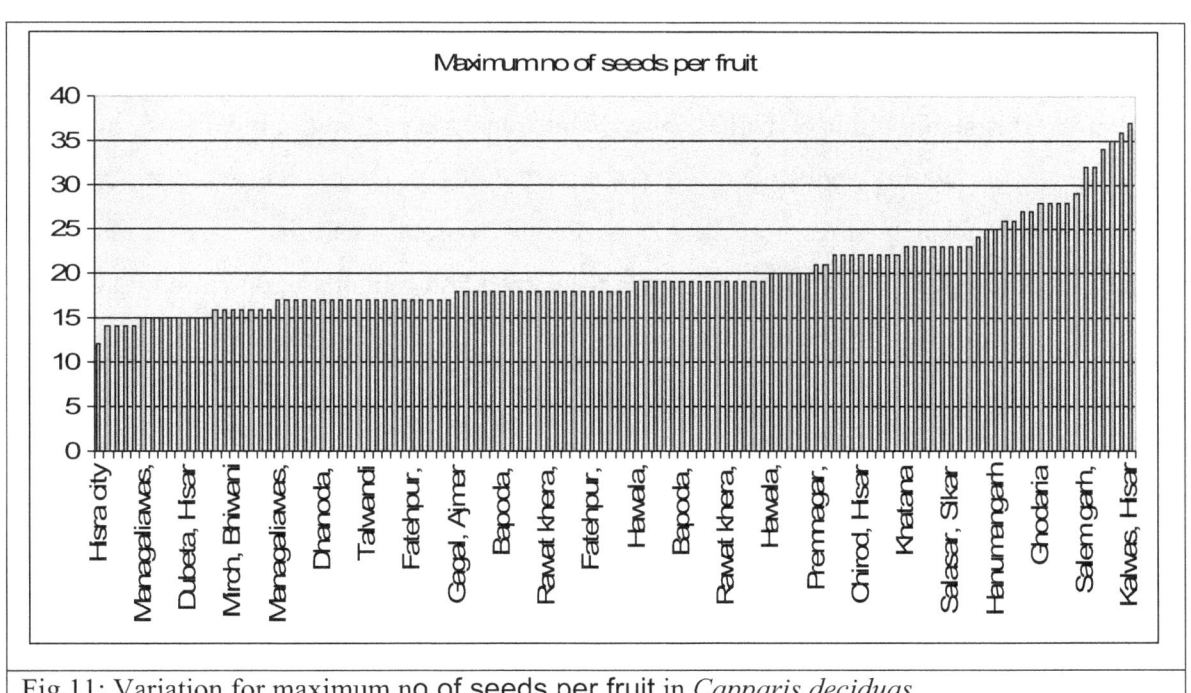

Fig.11: Variation for maximum no of seeds per fruit in *Capparis deciduas*

(xi) Difference for minimum and maximum no of seeds per fruit: Difference for minimum and maximum no of seeds per fruit varies (Fig 12) from 10 to 34 with a mean of 16.29. The highest value of difference for minimum and maximum no of seeds per fruit was more than three times of lowest value. The coefficient of variation for maximum no of seeds per fruit was 29.09 %. Most frequently occurring difference for minimum and maximum no of seeds per fruit were 17, 18 and 19.

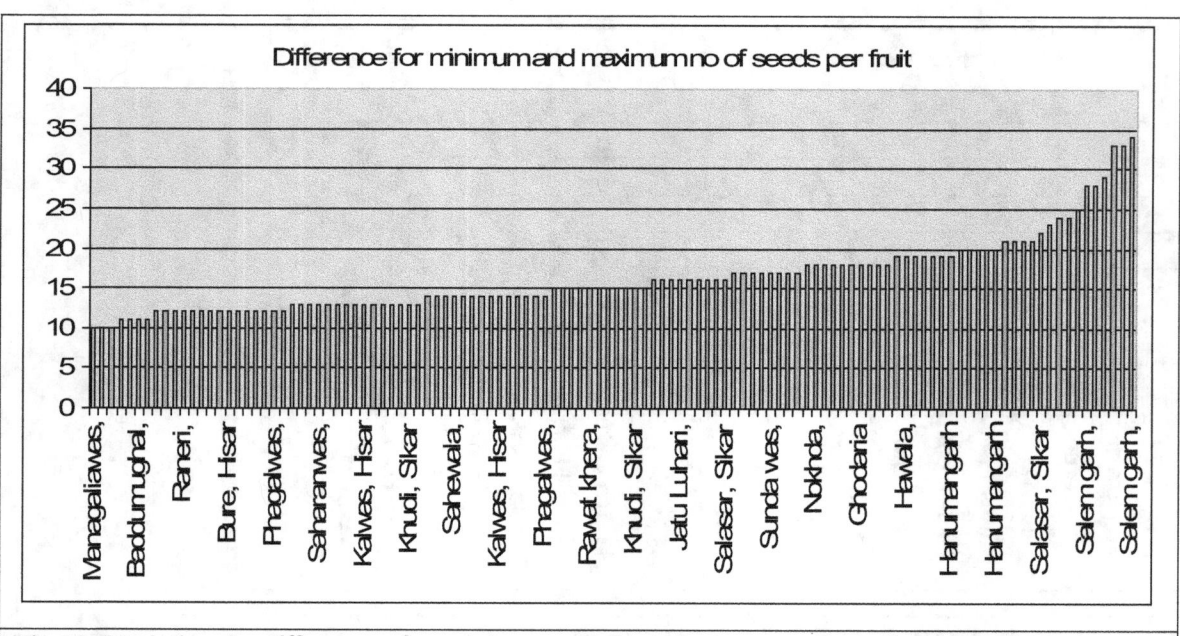

Fig.12: Variation for difference for minimum and maximum no of seeds per fruit in *Capparis decidua*

In natural population, rich genetic diversity with wide range of variability is available for plant types, bearing habit, fruit size, colour of fruits, spiny habit, plant spread and compactness of canopy, flower colour, time of flowering and fruiting, etc (Mahla et al 2010). In general, two distinct plant types of kair occur, tree form having more than 5 m height whereas majority occurred as bushes. It appears that plant attains tree form when it grows from seed and remains undisturbed. On the other hand, plants that get exposed to biotic interference may tend to produce more shoots and also propagates through root suckers. There are lot of variation exists for spine length (2-5 mm) but plants with very less rudimentary spines and sometimes spineless also found in nature. Kair flowers throughout the year; February - March *(Ambe Bahar),* July - August *(Mrig Bahar)* and October - November *(Hast Bahar)* but profuse flowering occurs only in *Ambe Bahar* which gives quality fruits in ample quantity. A wide diversity in flower colour can be seen from light red to scarlet red but plants with yellow flowers also exist in the natural stands of rangeland. There is very wide range in tender fruit yield per plant under natural stands (100 g to more than 5.0 kg) as it depends on biotic factors and grazing pressure under rangeland conditions.

(B) Documentation of traditional knowledge

Traditional Uses index of *Capparis decidua* was worked out in comparison to other trees/shrubs of arid region on the basis of views of local people (Table 2).

Table 2. Traditional Uses index of *Capparis decidua* in comparison to other trees

Sr no	Tree species	Food value	Medicinal value	Sand dune stabilization and salt tolerance	Average
1	*Capparis decidua*	0.92	0.91	0.90	0.91
2	*Azadirachta indica*	0.77	1.00	0.81	0.86
3	*Prosopis cineraria*	0.96	0.75	0.84	0.85
4	*Acacia nilotica*	0.83	0.79	0.90	0.84

Index score analysis on the basis of simultaneous consideration of medicinal value, food value and potential for sand dune stabilization and salt tolerance suggested the superiority of *Capparis decidua* (0.91) in comparison to *Azadirachta indica* (0.86), *Prosopis cineraria* (0.85) and *Acacia nilotica* (0.84). *Azadirachta indica* was found slightly better than Capparis decidua on the basis of medicinal importance. *Prosopis cineraria* was observed to be slightly better than *Capparis decidua* on the basis of food value. *Acacia nilotica* and *Capparis decidua* have equal potential for salt tolerance whereas *Capparis decidua* and *Prosopis cineraria* have equal potential for sandy region. Index score analysis on the basis of simultaneous consideration of medicinal value, food value and potential for sand dune stabilization and salt tolerance suggested the superiority of *Capparis decidua* in comparison to *Azadirachta indica*, *Prosopis cineraria* and *Acacia nilotica*. *Capparis decidua* and *Prosopis cineraria* have almost equal hardiness to bear extreme conditions of arid region. Moreover, *Capparis decidua* can be successfully grown readily from seed and root suckers. Seed production is a serious problem as both *Capparis decidua* and *Prosopis cineraria* are over exploited because of highly economic importance of their immature fruits.

- The flower buds and unripe green fruits of *Capparis decidua* are pickled and also cooked and eaten as vegetable. Pickle and cooked vegetables of unripe fruits are very useful for stomach troubles especially for constipation.

- The bark of *Capparis decidua* has been shown to be useful in the treatment of coughs, asthma, bronchial inflammation, indigestion and rheumatism. The stem bark decoction (10-15ml) is administered twice a day in asthma and other respiratory disorders. The stem bark is used as a laxative, diaphoretic and anthelmintic. The bark of its leafless shrub is used for the treatment of asthma, cough, inflammation and acute pain. The stem is used in pyorrhea and rheumatism. Powered coal of stem with water is taken for the treatment of fractured bone. The bark has an acrid, sharp, hot taste; analgesic, diaphoretic, alexeteric, laxative, in dropsy ground, anthelmintic; good in asthma, ulcers and boils, vomiting, piles and all inflammations.

- The root and root bark are pungent and bitter and are given to treat intermittent fevers and rheumatism. It is applied externally to ribs in case of pleurisy. The root bark extract is given twice a day for 3 days in the treatment of haemorrhoids. Charcoal of root is taken orally for rheumatism and bone fracture. The inner bark of roots is used to treat scabies and eczema. Root paste is applied on scorpion bite. Root bark and stem are reported to contain a spermidine alkaloid and isocodonocarpine effective in treatment of asthma, inflammation and cough (Ahmad *et al*, 1989).

- Paste of coal from wood is applied extremely to muscular injuries. Juice of fresh plant is dropped into the ear to kill worms. The tender branches and leaves are used as a plaster for boils and swellings and to relieve toothache on chewing. Paste made of aerial part is applied on fracture as an analgesic and anti-inflammatory. The top shoots and young leaves are made into a powder and used as a blister; they are also used in boils, eruptions and swellings and as an antidote to poison.

- The extract of unripe fruits and shoots are useful for controlling heart, liver and kidney problem. The fruits of *C. decidua* have medicinal value and are believed to provide relief from cardiac and gastric troubles. Many vaids prescribe ker fruits for cardiac trouble. The extract of immature fruits can be used to cure trachoma (Chronic conjunctivitis that can cause blindness). The powered fruit of *Capparis decidua* is used in anti-diabetic formulations. The fruit has a sharp hot astringent to the bowels; destroys foul breath, biliousness, and urinary purulent discharges; good in cardiac troubles. Floral primordial of *Capparis decidua*, boiled in oil and tied between the infected hooves, is found to be curative

The *Capparis decidua* has been traditionally useful in the treatment of coughs, asthma, bronchial inflammation, respiratory disorders, indigestion, pyorrhea, intermittent fevers, rheumatism, ulcers, boils, vomiting, piles, bone fracture inflammation and acute pain, scorpion bite, antidote to poison, toothache, diabetes, heart, liver and kidney problems, trachoma (Chronic conjunctivitis that can cause blindness), infected hooves, scabies and eczema. It is being used as laxative, tonic and mouth freshner. *Capparis decidua* is an indigenous medicinal plant, belongs to the family *Capparidaceae* and commonly known as 'Kurrel' in Hindi. It is a densely branching shrub with scanty, small, caducous leaves. Barks, leaves and roots of *C. decidua* have been claimed to relieve variety of ailments such as toothache, cough, asthma, intermittent fever and rheumatism (Dhar et al., 1972). The powdered fruit of *C. decidua* is used in anti-diabetic formulations (Yadav et al., 1997), while the bark of its leafless shrub is used for the treatment of asthma, cough, inflammation and acute pain (Ahmad et al., 1992). Seeds of *Capparis. decidua* showed antibacterial activity against *Vibriocholerae ogava*, *inaba* and *eltor* (Gaind et al., 1972). Joseph and Jini, 2011 reported that Capparis deciduas has significant pharmacological activities like hypercholesterolemic, anti-inflammatory and analgesic, antidiabeti, antimicrobial, antiplaque, antihypertensive, antihelmintic and purgativ activities. The female flowers of some of the *Capparis* species are used as vegetable and fruits are used in pickle production because of their high nutritive ingredients like proteins, carbohydrate, minerals and vitamins. Bhavna Sharma *et. al*, 2010 used the dried fruits of Capparis decidua as an ingredient in anti-diabetic compositions. The dried fruits are used as an ingredient in anti-diabetic compositions and the green berries are used in food preparations such as pickles owing to the ancient belief that it possesses medicinal properties (Yadav et. Al., 1997). Ground stems and leaves used in alveolaris and pyorrhea; Root bark is used as anthelmintic and purgative; Wood coal used in muscular injuries (Kirtikar and Basu, 1993; Chopra et al., 1999). Barks, leaves and roots of *Capparis decidua* have been claimed to relieve variety of ailments such as toothache, cough, asthma, intermittent fever and rheumatism (Dhar et al., 1972). The powdered fruit of *Capparis decidua* is used in anti-diabetic formulations (Yadav et al., 1997a, b), while the bark of its leafless shrub is used for the treatment of asthma, cough, inflammation and acute pain (Ahmad et al., 1992). Capparis decidua powder is used against alloxan induced oxidative stress and diabetes in rats (Agarwal and Chauhan, 1988; Yadav et al., 1997). The aqueous extracts of roots of *Capparis decidua* are found to have purgative activity (Gaind et al., 1969) while the alcoholic extract of the fruit pulp and root bark possess anthelmintic activity (Gaind et al., 1969; Mali et al., 2004). *Capparis decidua* fruit and flower extract have potent activity in preventing plaque formation (Rathee et al., 2010). The extract of unripe fruits and shoots of Capparis decidua cause reduction in plasma triglycerides, total lipids and phospholipids; hence used as hypercholesterolemic. It appeared to operate through increased fecal excretion of cholesterol as well as bile acids (Goyal and Grewal, 2003). The serum cholesterol level was reduced by 61, 58, 48 and 28% in *Capparis decidua* fruit, flower, shoot and bark after a dose of 500 mg kg-1 b.wt. was given to rabbits (Purohit and Vyas, 2005, 2006; Sharma et al., 1991). The extracts of *Capparis decidua* prove to have a hypolipidemic potential (Chahlia, 2009). Capparis decidua has considerable nutritional value. Fruit is a rich source of vitamin C (Chauhan et al., 1986;

Duhan et al., 1992). The presence of oil content in seed and flower along with sugar and protein substantiate the nutritional value (Rai and Rai, 1987). According to ethnopharmacological relevance the dried fruits of *Capparis decidua* are used as ingredients in antidiabetic composition (Shikha et al 2011). The bark of *Capparis decidua* is bitter and diuretic. It is given in hepatic, spleen and renal complaints. It is used as anti-inflammatory used for enlarged cervical glands, sciatica, rheumatoid arthritis, externally on swelling. Fruits and seeds are used for urinary purulent discharges and dysentery and antimicrobial (Sharma et al; 2011). *Capparis decidua,* climbing, thorny shrub, densely branched, spinous shrub or tree, up to 6 meters in height, is widely used in traditional medicinal system of India, has been reported to possess carminative, tonic, emmenagogue, aphrodisiac, alexipharmic; improves the appetite, antirheumatic, lumbago, hiccough, cough and asthma. It is known as a rich source of alkaloids, phenols, sterols and glycosides (Singh et al. 2011). The innumerable medicinal properties and therapeutic uses of *Capparis decidua* as well as its phytochemical investigations prove its importance as a valuable medicinal plant.

- *Capparis decidua* provides hard, heavy, termite resistant/timber. White ants do not attack the ker wood; therefore, it can even be used to make agricultural implements like plough and the yoke. Ker wood being very strong and durable is used to make the foundation around the well and firewood. Being non shrinkage nature of wood, it is specifically used as central liver (kila and mani) in household chaki (plate 9).

Plate 9: Wood of *Capparis decidua* as a central liver (kila and mani) in household chaki

- Exports and prices of *Capparis decidua* (young fruits), *'Sangari'* (young pods of *Prosopis cineraria*), *'Kumbat'* (seeds of *Acacia Senegal*), *and 'Gawar'* (immature dried pods of *Cyamopsistetragonoloba*) have rocketed. The diversity, which had been preserved and maintained by the communities, has become a source of exploitation. The material they have preserved through centuries of sustenance has gone to the hands of commercial entrepreneurs.

- *In a guest post* Ms. Vishnu Priya *shares the recipe for preparing Ker Sangari Pickle.* Ker is a green berry like fruit of a thorny bush (*capparis decidua*) found in the arid parts of Rajastnan. It is very sour, and is commonly available during the months of April and May. In Jaipur everyone can buy fresh ker from Choti Choupar. The traders go to the extent of sorting ker by size as the smaller berries are considered to be tastier. Local Pansari shops in the walled city provide dried ker all the year round. Fresh Ker has to be put in an earthen utensil or preferably a matka in salt water for 15 days so that it loses its sourness. After two weeks the salt water is thrown away and the vegetable is dried in shade to be stored for the rest of the year or to be cooked right then. Because of its numerous medicinal qualities ker is also used in many ayurvedic medicines.

- Sangari are the bean like fruit of the Khejari (*prosopis cineraria*) tree. This tree is commonly found in rajasthan. And it gives its fruit, sangari, in the winter's months. Thin and kachi sangri are plucked from the tree and lightly boiled and dried in shade to be used during rest of the year. They can be eaten as a pickle, vegetable or a snack-fry in oil & serve with salt and red chillies. Ingredients:150 gm Dry Sangari, 100 gm Dry Ker, 50 gm Dry Lasore, 250 gm Sesame Oil (tilli ka tel) OR Mustard Oil(Sarson Ka Tel), 50 gm Aniseed Powder (Saunf), 50 gm Kachri Powder (A flavoring agent commonly found in Rajasthan.Also used to make uncooked meat tender), 25 gm pomegranate seed powder (Anar dana), 100 gm Mustard seed powder(Raai), 1/4 Tea Spoon Asafoetida Powder (Heeng), 1 Tea Spoon Mustard Seeds (Sarson), 1 Tea Spoon Cumin Seeds (Jeera), 1/2 Tea Spoon Fenugreek Seeds (Methi), 1 Tea Spoon Salt, 1 Tea spoon Red Chillies Powder, 2 Tea spoon Turmeric.

- Method: Soak the dried Ker, Sangri & Lasore in water overnight. Throw away the water the vegetables were soaked in overnight and add the vegetables to a pot of boiling salt water till they are soft. Throw away the salt water and put out the vegetables to dry on a soft cloth. Heat the oil in a Wok till it is smoking and crackle cumin, mustard, fenugreek seeds and add asafoetida and take the wok off the flame. Add all ingredients except raai and the vegetable and cook the masala on low flame till the masala is cooked. Let the masala cool down after it is cooked and add the vegetables and raai and mix well and store in a glass jar for 2 to 3 days stirring 2-3 times daily till the mixture turns sour-khatta. The pickle is ready to be served with poories or parathas. If the pickle is to be stored for a long period of time, take enough oil to cover the pickle in the jar. Heat it till it is smoking. Let it cool down. Then add it to pickle in the Jar so that the pickle is covered in oil. The oil will make the pickle last longer. The commercial packing of pickle is shown in plate 10.

Plate 10: Pickle of *Capparis decidua* with its commercial packing

- Post harvest technology: The unripe/ripe fruits of kair are generally not eaten fresh due to their acrid taste, but can be converted into a variety of by-products after processing. The pickles are the most commonly and widely utilized post harvest product of *Capparis decidua*. The processed fruits can be utilized directly for preparation of pickles or as a vegetable or can be dehydrated for off-season Utilization Based on the size, three relative grades of processed *kair* are available in the market; big, medium and small size. In fact, the basis of size grading is the relative maturity of fruits. The smaller fruits are more tender and of better quality than the bigger fruits. The processed fruits are stored either in pots or in plastic containers while processed dried fruits are stored in flexible polybags. The dried fruits can be stored in polybags for a year without any deterioration in the quality.

- Livelihood dynamics around *Capparis decidua*: *Capparis decidua* is of much use in climate prediction and features in farmers' strategies in natural resources production and management and agricultural planning. The local farmers of Surender Nagar, Kheda, Bhavnagar and Ahmedabad of Gujarat state that they use kair (locally called kerda) in predicting the weather, namely temperature and rainfall. Apart from the immense use of kerda in making local pickle, the farmers consider that if blooming in kerda is greater and flowers are deep pink, then the temperature is more than 45°C and the rainfall will be less than normal. Based on the observations of the numbers of flowers and fruits and the canopy of kerda, the farmers select their crop varieties and cropping systems for the following rainy season. The kerda pickle is made by poor local farmers and sold in local markets to generate income. It is also used in ethnomedicine, including treating stomach disorders and skin diseases, for both humans and animals.

- Religious value: *Capparis decidua* is often found associated with religious rituals of the local populations of the Thar Desert. It can easily be seen growing near or around temples and other religious places. It is a custom for newly married couples and the newly born child to offer prayers in front of a kair plant. This plant is also associated with graves and crematoriums as it is considered holy.

(C) Identification of promising reproductive material of *Capparis decidua* from different fragile environmental conditions

A survey of southern Haryana and Rajasthan was conducted during flowering and fruiting seasons. Large sized trees up to 8.5 metres (Plate 11) height with higher yield of fruit per plant were observed in Salasar, Sikar (Rajasthan), Kalwas, Hisar (Haryana) and Balsamand, Hisar (Haryana). In general, two distinct plant types of kair occur: a tree form, which is relatively unusual, and a shrub form, in the majority of plants (Singh and Singh, 2011). It appears that tree form is attained when the plant grows from seed and remains undisturbed. On the other hand, plants exposed to biotic interference may tend to produce more shoots and also propagate through root suckers. This view is supported by the fact that *C. decidua* occurred singly in tree form and mostly in clusters in bush form. The plant can also be trained by allowing a single stem to grow, and can be integrated as one of the components in different cropping models. However, scientific evaluation is needed to assess the relative compatibility and economic viability under different resource situations. Plants mostly shrubs with large spread and higher yield of fruit per plant were obsereved in Sahewala, Hisar (Haryana), Bapoda, Bhiwani (Haryana), Fatehpur, Sikar (Rajasthan), Salasar, Sikar (Rajasthan), Ghodaria Khurad, Sikar (Rajasthan), Nokhda, Bikaner (Rajasthan), Phagalwas, Sikar (Rajasthan) and Dadar, Churu (Rajasthan). Progenies of these superior mother trees/shrubs/plants have been established in the germplasm bank.

Plate 11a: Tree form of *Capparis decidua*

Plate 11b: Tree form of *Capparis decidua*

Plate 11c: Tree form of *Capparis decidua*

Plate 11d: Tree form of *Capparis decidua*

Plate 11e: Tree form of *Capparis decidua*

Plate 11f: Tree form of *Capparis decidua*

Plate 11g: Tree form of *Capparis decidua*

(D) Establishment of Germplasm Bank

(a) 1st Set of Germplasm Bank

(i) Raising of Seedlings: The seeds from nine mother treesshrubs/plants of *Capparis decidua* growing near Hisar were collected in September, 2008 prior to the start of project. Fifty seeds of each of nine progenies were sown in polythene bags for raising seedlings during February 2009. These seedlings attain the height of 10-15 cm by July 2009.

(ii) Field Plantation: Seedlings of nine progenies were transplanted in the field following Randomized Block Design (RBD) with three replications during September, 2009. Data recorded for plant height (cm), plant spread (cm) and root length (cm) in December, 2010 (fifteen months after transplanting) are presented in Table 3. The plant height varies from 31.2 cm to 43.5 cm with a mean of 37.5 cm whereas plant spread varies from 27.5 cm to 39.4 cm with a mean of 35.8 cm. The root length was observed higher than plant height and plant spread (Fig. 13). The root length varies from 39.4 cm to 86.5 cm with a mean of 65 cm. Root length varies from 1.26 times of plant height to 2.04 times of plant height with a mean of 1.73 times of plant height whereas root length varies from 1.43 times of plant spread to 2.06 times of plant spread with a mean of 1.81 times of plant spread. Coefficient of variation for root length was 25.78 per cent whereas for plant height and plant spread, coefficients of variations were 11.20 per cent and 16.53 per cent, respectively. Plants along with roots are shown in Plate 12.

Table 3: Plant height, plant spread and root length of nine progenies of *Capparis decidua* at the age of 15 months after transplanting

Progenies	Plant Height (cm)	Plant Spread (cm)	Root length (cm)
P-1	36.4	36.4	74.3
P-2	35.9	36.7	71.6
P-3	43.5	39.4	81.2
P-4	34.8	32.4	52.5
P-5	42.9	45.4	86.5
P-6	38.6	38.2	59.6
P-7	40.2	37.4	75.7
P-8	31.2	27.5	39.4
P-9	33.8	29.1	44.5
Mean	37.5	35.8	65.0
CD (5% level of significance	1.19	1.49	3.01
Range	31.2-43.5	27.5-39.4	39.4-86.5
SD	4.20	5.92	16.69
CV	11.20	16.53	25.78

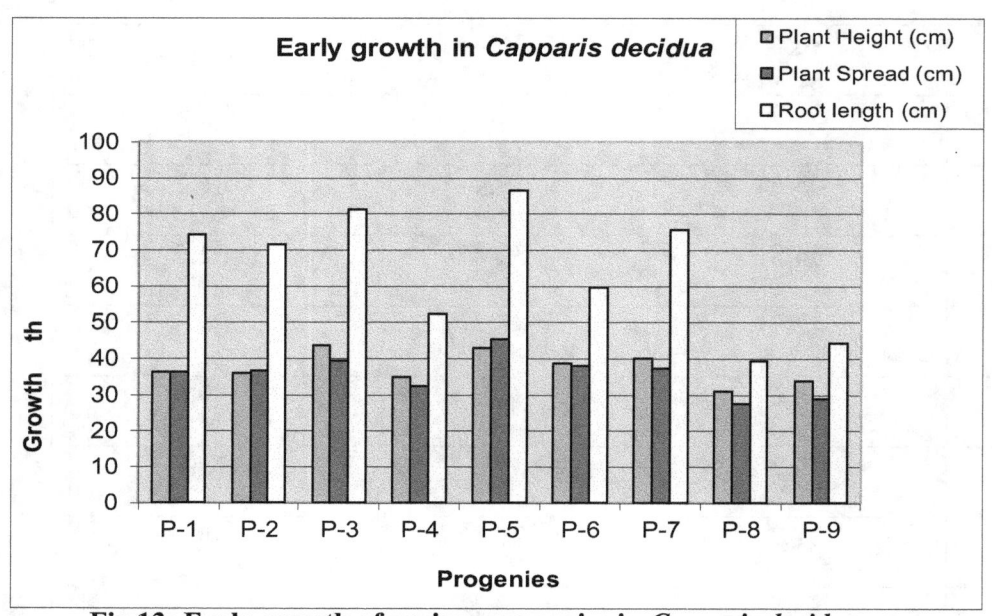

Fig.13: Early growth of various progenies in *Capparis decidua*

Plate 12: Plants of *Capparis decidua* along with roots

(b) 2nd Set of Germplasm Bank

(i) Raising of Seedlings: Matured fruits of *Capparis decidua* were collected from fifty-five plants/shrubs/trees growing naturally in southern Haryana, adjoining Rajasthan and Bikaner District of Rajasthan during May 2009. Fifty seeds of each of fifty-five accessions recorded as 1-55 were sown in polythene bags during early August, 2009 for raising seedlings (Seedling stage shown in plate 13).

(ii) Field Plantation: Seedlings of thirty-nine accessions collected from southern Haryana, adjoining Rajasthan and Bikaner region of Rajasthan were transplanted in the field following Randomized Block Design (RBD) with three replications during February, 2010. The data recorded in the end of November, 2010 and is presented in Table 4. The highest plant height of 24 cm was observed in accession no 51 from Kalwas, Hisar (Haryana) followed by accession nos 17 and 16 from Nokhda, Bikaner (Rajasthan), accession no 40 from Bapoda, Bhiwani (Haryana), accession no 28 from Sahewala, Hisar (Haryana) and accession no 4 from Hawala, Rajsamand (Rajasthan). The progeny no 51 from Kalwas, Hisar (Haryana) and progeny no 28 from Sahewala, Hisar (Haryana) were observed comparatively more erect growing (Plate 14). The plant height at the age of nine months after transplanting varies from 13.4 cm in accession no 33 from Loharu, Bhiwani (Haryana) to 24.0 cm in accession no 51 from Kalwas, Hisar (Haryana) with a mean of 19.56 cm and 14.41 per cent coefficient of variation.

Table 4: Plant Height of 39 Germplasm Accessions of *Capparis decidua* collected from southern Haryana, adjoining Rajasthan and Bikaner region of Rajasthan at the age of nine months after transplanting

Accession no	Location	Plant Height (cm)
1	Khatana Khera, Ajmer (Rajasthan)	21.6
2	Khatana Khera, Ajmer (Rajasthan)	20.8
4	Hawala, Rajsamand (Rajasthan)	22.9
6	Hawala, Rajsamand (Rajasthan)	22.1
8	Palukalam, Jaipur (Rajasthan)	17.3
9	Hawala, Rajsamand (Rajasthan)	20.6
12	Gajner, Bikaner (Rajasthan)	18.4
13	Kolayat, Bikaner (Rajasthan)	19.1
14	Raneri, Bikaner (Rajasthan)	16.3
15	Raneri, Bikaner (Rajasthan)	16.6
16	Nokhda, Bikaner (Rajasthan)	23.5
17	Nokhda, Bikaner (Rajasthan)	23.8
18	Nokhda, Bikaner (Rajasthan)	21.7
19	Bhana, Bikaner (Rajasthan)	17.2
20	Raneri, Bikaner (Rajasthan)	21.5
22	Raneri, Bikaner (Rajasthan)	21.0
23	Raneri, Bikaner (Rajasthan)	20.3
24	Raneri, Bikaner (Rajasthan)	18.6
28	Sahewala, Hisar (Haryana)	23.0
31	Kharkarhi, Bhiwani (Haryana)	19.2
32	Dhani Tohae, Bhiwani (Haryana)	16.8
33	Loharu, Bhiwani (Haryana)	13.4
34	Madhogarh, Mohindergarh (Haryana)	21.2
35	Dhanoda, Mohindergarh (Haryana)	20.7

36	Dhanoda, Mohindergarh (Haryana)	19.2
37	Saharanwas, Rewari (Haryana)	15.5
40	Bapoda, Bhiwani (Haryana)	23.3
41	Bapoda, Bhiwani (Haryana)	20.9
42	Premnagar, Bhiwani (Haryana)	19.4
43	Jatu Luhari, Bhiwani (Haryana)	16.8
44	Jatu Luhari, Bhiwani (Haryana)	21.4
46	Rawat khera, Hisar (Haryana)	15.5
49	Chirod, Hisar (Haryana)	17.9
50	Chirod, Hisar (Haryana)	16.1
51	Kalwas, Hisar (Haryana)	24.0
52	Salem garh, Hisar (Haryana)	17.4
53	Salem garh, Hisar (Haryana)	14.9
54	Sunda was, Hisar (Haryana)	21.9
55	Sunda was, Hisar (Haryana)	21.2
Mean		**19.56**
CD (5% level of significance)		**0.99**
Range		**13.4-24.0**
Standard deviation		**2.81**
Coefficient of Variation		**14.41**

Plate 13: Seedling Stage of *Capparis decidua*

Plate 14: *Erect growing plants of Capparis decidua* **in field plantation**

(c) 3rd Set of Germplasm Bank

(i) Raising of Seedlings: Again, during October-November 2009, matured fruits were collected from 61 plants/shrubs/trees growing naturally in southern Haryana and adjoining Rajasthan. Fifty seeds of each of 116 accessions collected in total (seeds collected from fifty-five plants/shrubs/trees growing naturally in southern Haryana, adjoining Rajasthan and Bikaner District of Rajasthan during May 2009 recorded as 1-55 + seeds collected from 61 plants/shrubs/trees growing naturally in southern Haryana and adjoining Rajasthan recorded as 101-161) were sown in polythene bags during early February, 2010 for raising seedlings.

(ii) Field Plantation: Seedlings of seventy-seven accessions collected from southern Haryana, adjoining Rajasthan and Bikaner region of Rajasthan were transplanted in the field following Randomized Block Design (RBD) with three replications during September, 2010. The data recorded in the early December, 2010 is presented in Table 5. The highest plant height of 20.6 cm was observed in accession no 142 from Salasar, Sikar (Rajasthan) followed by accession no 51 from Kalwas, Hisar (Haryana), accession no 40 from Bapoda, Bhiwani (Haryana), accession no 16 from Nokhda, Bikaner (Rajasthan), accession no 118 from Nokhda, Bikaner (Rajasthan) and accession no 128 from Fatehpur, Sikar (Rajasthan). The plant height (plate 15) at the age of three months after transplanting varies from 11.4 cm in accession no 33 from Loharu, Bhiwani (Haryana) to 20.6 cm in accession no 142 from Salasar, Sikar (Rajasthan) with a mean of 15.52 cm and 13.91 per cent coefficient of variation.

Table 5: Plant height of 77 germplasm accessions of Capparis decidua collected from southern Haryana, adjoining Rajasthan and Bikaner region of Rajasthan at the age of three months after transplanting

Accession no	Location	Plant Height (cm)
2	Khatana Khera, Ajmer (Rajasthan)	15.4
4	Hawala, Rajsamand (Rajasthan)	17.5
8	Palukalam, Jaipur (Rajasthan)	13.6
9	Hawala, Rajsamand (Rajasthan)	15.2
12	Gajner, Bikaner (Rajasthan)	13.4
13	Kolayat, Bikaner (Rajasthan)	14.2
14	Raneri, Bikaner (Rajasthan)	12.9
15	Raneri, Bikaner (Rajasthan)	12.4
16	Nokhda, Bikaner (Rajasthan)	19.1
17	Nokhda, Bikaner (Rajasthan)	17.8
19	Bhana, Bikaner (Rajasthan)	12.7
20	Raneri, Bikaner (Rajasthan)	16.2
21	Raneri, Bikaner (Rajasthan)	15.6
22	Raneri, Bikaner (Rajasthan)	15.8
23	Raneri, Bikaner (Rajasthan)	15.6
24	Raneri, Bikaner (Rajasthan)	13.2
31	Kharkarhi, Bhiwani (Haryana)	14.4
32	Dhani Tohae, Bhiwani (Haryana)	12.3
33	Loharu, Bhiwani (Haryana)	11.4
35	Dhanoda, Mohindergarh (Haryana)	15.3
36	Dhanoda, Mohindergarh (Haryana)	14
37	Saharanwas, Rewari (Haryana)	12.7

40	Bapoda, Bhiwani (Haryana)	19.2
41	Bapoda, Bhiwani (Haryana)	16.5
42	Premnagar, Bhiwani (Haryana)	14.2
43	Jatu Luhari, Bhiwani (Haryana)	13.2
44	Jatu Luhari, Bhiwani (Haryana)	16.2
50	Chirod, Hisar (Haryana)	12.5
51	Kalwas, Hisar (Haryana)	20.2
52	Salem garh, Hisar (Haryana)	12.9
53	Salem garh, Hisar (Haryana)	12
54	Sunda was, Hisar (Haryana)	16.9
55	Sunda was, Hisar (Haryana)	16.5
101	Arya nagar, Hisar (Haryana)	15.8
102	Arya nagar, Hisar (Haryana)	15
103	Hisra city (Haryana)	15.1
104	Khedar, Hisra (Haryana)	16.9
105	Rawat khera, Hisar (Haryana)	17.5
106	Rawat khera, Hisar (Haryana)	16
107	Rawat khera, Hisar (Haryana)	14.7
108	Rawat khera, Hisar (Haryana)	13.9
109	Rawat khera, Hisar (Haryana)	17.2
111	Kalwas, Hisar (Haryana)	18
112	Kalwas, Hisar (Haryana)	18.3
113	Kalwas, Hisar (Haryana)	17.3
116	Talwandi Ruka, Hisar (Haryana)	12.7
117	Bure, Hisar (Haryana)	13.1

118	Dubeta, Hisar (Haryana)	19
120	Dadar, Churu (Rajasthan)	14.6
123	Dadar, Churu (Rajasthan)	15.4
124	Dadar, Churu (Rajasthan)	15.9
128	Fatehpur, Sikar (Rajasthan)	18.7
130	Fatehpur, Sikar (Rajasthan)	18
131	Fatehpur, Sikar (Rajasthan)	18.2
132	Ghashu Madhopura, Sikar (Rajasthan)	16.4
133	Ghashu Madhopura, Sikar (Rajasthan)	15.5
134	Khudi, Sikar (Rajasthan)	14.7
135	Khudi, Sikar (Rajasthan)	15.2
136	Phagalwas, Sikar (Rajasthan)	17
137	Phagalwas, Sikar (Rajasthan)	16.5
138	Phagalwas, Sikar (Rajasthan)	15.8
140	Phagalwas, Sikar (Rajasthan)	16.8
142	Salasar, Sikar (Rajasthan)	20.6
144	Salasar, Sikar (Rajasthan)	18.2
145	Salasar, Sikar (Rajasthan)	17.7
149	Ghodaria Khurad, Sikar (Rajasthan)	16.4
150	Ghodaria Khurad, Sikar (Rajasthan)	15.5
151	Ghodaria Khurad, Sikar (Rajasthan)	12.4
152	Ghodaria Khurad, Sikar (Rajasthan)	13.3
153	Chirawa, Jhunjhanu (Rajasthan)	14.8
154	Badi-sahi, Jhunjhanu (Rajasthan)	15.4
155	Shailu, Jhunjhanu (Rajasthan)	17.5

156	Hanumangarh (Rajasthan)	15.4
157	Hanumangarh (Rajasthan)	16.6
158	Hanumangarh (Rajasthan)	12.2
159	Hanumangarh (Rajasthan)	13.7
160	Hanumangarh (Rajasthan)	15.4
Mean		15.52
CD (5% level of significance)		**0.97**
Range		11.4-20.6
Standard deviation		2.10
Coefficient of Variation		13.91

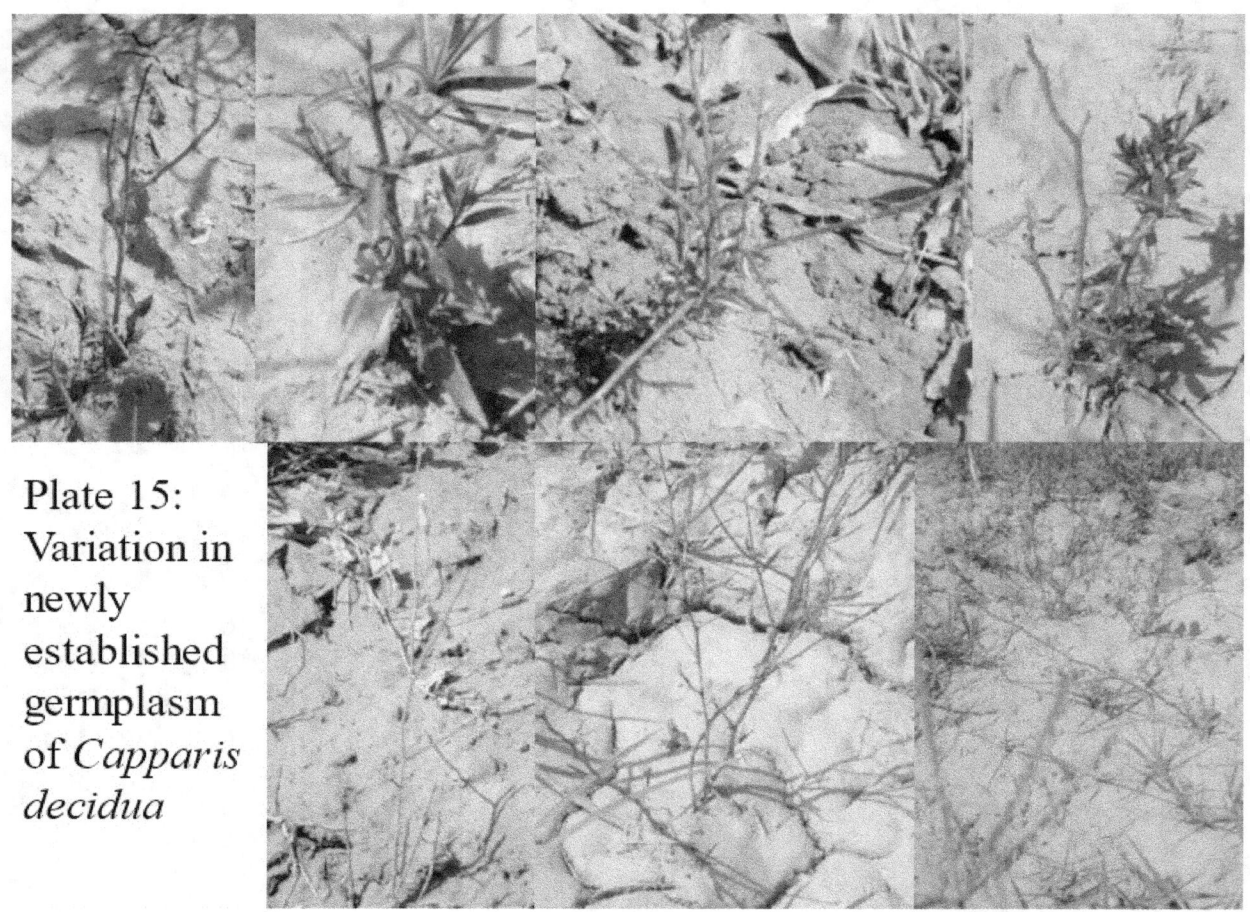

Plate 15: Variation in newly established germplasm of *Capparis decidua*

(E) Domestication and Breeding of *Capparis decidua*

Reproductive biology of *Capparis decidua* with respect to seed production age, flower development, breeding system, seed quality and seed storability were worked out. Plus trees will be selected from different regions and progeny test will be established for identification of genetically superior mother plants for large-scale production of genetically superior seed and other vegetatively propagating material. Package of practices of this species will be developed for large-scale plantation.

Flowering and fruiting in *Capparis decidua* have been observed twice (March-April to June and August to October-November) in a year with a seed collection time in May-June and October-November. Lot of variation was observed for flowering and fruiting between regions and between plants with in region. It was observed that some of the plants/shrubs/trees of *Capparis decidua* have one time flowering and fruiting in a year whereas others have two times flowering and fruiting in a year. Very few plants have alternate year flowering and fruiting. From a group of *Capparis decidua* plants/shrubs/trees at a particular place, some plants were observed to have full flowering whereas other plants were observed to have flowering ranging from zero to 80 %. Variation has been observed for flower colour. Variability in flower colour in natural stands of *Capparis decidua* was reported with brick red- and yellow-coloured flowers (Singh and Singh, 2011). In the present scenario, where there is no commercial cultivation of this species, and green fruits are harvested by the rural people, it is obvious that more fruits will be picked up from the non-spiny open type shrubs that are earlier fruiting. The flowers of *Capparis decidua* were observed to be complete having all the four parts (Plate 5) viz., calyx (four sepals), corolla (four petals), androecium (many stamens usually 13-16 with a length varies from 0.5 to 1.2 cm) and gynoecium (ovary superior, bicarpellary syncarpous, unilocular with many ovules arranged on parental placenta, style absent and stigma about 1.1 cm.). In *Capparis decidua* stigma is closely surrounded by anthers. Pollination may be before flower opening or after flower opening. But the position of anthers in relation to stigma seems to ensure self-pollination.

Collection of matured seed is a serious problem in *Capparis decidua* because its unripe fruits are in great demand for making pickle at commercially large scale. Fruits of *Capparis decidua* are also in great demand for medicinal uses. Poor landless people used to collect almost all the unripe fruits as these plants/shrubs/trees are naturally growing in undisturbed common lands. The fruits when turned pink are to be collected for the purpose of seeds collection. The fruits are to be washed and seeds are to be separated from the pulp. Fruits were collected at different stages of maturity. The best stage of fruit collection on the basis of results of germination was that when fruit turned pink. The retention of seed viability in *Capparis decidua* was observed to have negative correlation with duration of period from fruit collection to separation of seed from pulp. The retention of seed viability in *Capparis decidua* was observed up to one year where seeds were separated from pulp of fruit immediately after fruit collection from tree/shrub/plant.

The seeds of *Capparis decidua* showed lot of variation for seed germination (fresh seeds) ranging from zero percent to 63.89%. About fifty percent seed lots showed increase in germination percent up to three months after storage. This is clear indication of presence of dormancy in different seed lots of *Capparis decidua*. Four seed lots from diverse regions showed increase in germination percent up to six months consistently. These observations clearly suggested the extent of variation for seed dormancy among different seed lots of *Capparis decidua*.

A survey of southern Haryana and Rajasthan was conducted during flowering and fruiting seasons. Large sized trees up to 8.5 metres height with higher yield of fruit per plant were observed in Salasar, Sikar (Rajasthan), Kalwas, Hisar (Haryana) and Balsamand, Hisar (Haryana). Plants mostly shrubs with large spread and higher yield of fruit per plant were obsereved in Sahewala, Hisar (Haryana), Bapoda, Bhiwani (Haryana), Fatehpur, Sikar (Rajasthan), Salasar, Sikar (Rajasthan), Ghodaria Khurad, Sikar (Rajasthan), Nokhda, Bikaner (Rajasthan), Phagalwas, Sikar (Rajasthan) and Dadar, Churu (Rajasthan). Progenies of these superior mother trees/shrubs/plants have been established in the germplasm bank. The germplasm bank of *Capparis decidua* will provide opportunity for the identification of genetically superior reproductive material for domestication this valuable tree species.

In order to standardized the production technology of *Capparis decidua*, the seeds were collected in different months, seeds were sown in different seasons, seed sowing was done in different composition of FYM and planting was done in different months. The root suckers of *Capparis decidua* were planted round the year with an interval of one month. It has been established that *Capparis decidua* can be successfully grown from seeds and suckers. Its matured fruits which have turned pink should be collected either in May-June or August-October. Seeds should be separated out by washing the pulp of fruits with in 2-3 days. Seeds are to be sown in polythene bags (with 3:1 ratio of sandy soil and FYM) in August if seeds are collected in May-June or February if seeds are collected in August-October. Collection of seeds in August-October and sowing of seeds in early February are desirable. Transplanting of seedlings should be done in early February by choosing seedlings of more than 15 cm height. Suckers of *Capparis decidua* are to be collected in early February and should be established in polythene bags (with 3:1 ratio of sandy soil and FYM). Plants raised from suckers attain usually attain a height of more than 15 cm in 4-5 months and hence these plants are transplanted in field during rainy season.

Tyagi et. al. 2010 developed protocol for *in vitro* multiplication of *Capparis decidua* (Forsk.) Edgew from cultured leaves procured from multiplying axillary shoots on the cultured nodal explants. The highest efficiency of shoot formation was observed on Murashige and Skoog (MS) medium containing 2 mg dm-3 benzyladenine (BA) and 0.5 mg dm-3 1-naphthaleneacetic acid. The regenerated shoots were transferred to MS medium containing 3 mg dm-3 BA for growth and proliferation. Shoots above 2 cm in length were transferred to MS medium supplemented with 1 mg dm-3 indole-3- butyric acid plus 0.5 mg dm-3 indole-3-acetic acid for root induction. Chahar et al., 2010 standardized the protocol for clonal propagation in *Capparis decidua*. They reported 96% shoot induction response on MS basal medium supplemented with 5.0 mg/L BAP and 25.0 mg/L adenine sulphate (MS17) with 3-4 shoots/explants in case of nodal explants. Shoot tips gave 25% shoot regeneration response on the same medium with 1-2 shoots/explants.

Bhargava et al., 2006 investigated the propagation behavior of *Capparis decidua*. They reported that semi-hard wood cuttings showed 40% sprouting and 20% rooting. They further reported that January, March and August to November are the suitable months for taking cuttings and IBA (7500 ppm) + Thymine (1000 ppm), prepared in Dimethyl Sulphoxide (DMSO) medium, was found to be optimum concentration.

Likely Impact of the Work on the Scientific Potential of Our Country

Poor or lack of seed production continues to be the major cause for *kair's* declining population. In fact, the plant has even been listed as one of the endangered species in India (Singh and Singh, 2011). It is also Suppressed by *Prosopis juliflora,* which regenerates faster and grows aggressively, but unlike kair, cannot be used as fodder. Hence, *Prosopis juliflora* is seriously threatening the survival of *Capparis decidua* and other indigenous species.

- The outcome of study (The flowers of *Capparis decidua* were observed to be complete having all the four parts viz., calyx, corolla, androecium and gynoecium. In *Capparis decidua* stigma is closely surrounded by anthers. But the position of anthers in relation to stigma seems to ensure self-pollination. *Capparis decidua is* predominantly self-pollinating species.) will be useful for further breeding programme and multiplication of superior reproductive material of *Capparis decidua.*

- The traditional superiority of *Capparis decidua* in comparison to *Azadirachta indica*, *Prosopis cineraria* and *Acacia nilotica* on the basis of simultaneous consideration of medicinal value, food value and potential for sand dune stabilization and salt tolerance will attract researchers, environmentalists, health specialists and businessmen for concentrating their efforts towards *Capparis deciduas.*

- Traditional knowledge of *Capparis decidua* being used in climate prediction will be standardized scientifically for prediction of climate.

- The superior genetic stock (The large sized trees up to 8.5 metres with higher yield of fruit per plant were observed in Salasar, Sikar in Rajasthan, Kalwas, Hisar in Haryana and Balsamand, Hisar in Haryana) of *Capparis deciduas.*

- Superior progenies from progeny testing of *Capparis decidua* will be useful for further multiplication of superior reproductive material.

- suggested lot of variation for above ground and below ground growth and the root length was observed higher than plant height and plant spread. Genetic control of erectness in plant growth of *Capparis decidua* will be useful for breeding *Capparis decidua* for straight growing plant type.

- The presence of lot of variation in germplasm of *Capparis decidua* will be useful for identification of genetically superior reproductive material of *Capparis decidua*.

- The cultivation practice (It has been established that *Capparis decidua* can be successfully grown from seeds and suckers. Its matured fruits which have turned pink should be collected either in May-June or August-October. Seeds should be separated out by washing the pulp of fruits with in 2-3 days. Seeds are to be sown in polythene bags (with 3:1 ratio of sandy soil and FYM) in August if seeds are collected in May-June or February if seeds are collected in August-October. Collection of seeds in August-October and sowing of seeds in early February are desirable. Transplanting of seedlings should be done in early February by choosing seedlings of more than 15 cm height. Suckers of *Capparis decidua* are to be collected in early February and should be established in polythene bags (with 3:1 ratio of sandy soil and FYM). Plants raised from suckers attain usually attain a height of more than 15 cm in 4-5 months and hence these plants are transplanted in field) will provide a great opportunity for farmers of arid zone for growing *Capparis decidua* on their farm for handsome income and to provide way for domestication of valuable endangered tree/shrub/plant.

Bibliography

1. Abdel-Mawgood, AL; Assaeed, AM and Al-Abdallatif, TI 2005. Genetic Diversity in an isolated population of Capparis decidua. FAO international workshop "the role of biotechnology for characterization and conservation of crop, forestry, animal and fishery genetic resources". Torino, Italy March 6-8, 2005.

2. Abdel-Mawgood, AL; Assaeed, AM and Al-Abdallatif, TI 2006. Application of RAPD Technique for the Conservation of an Isolated Population of Capparis decidua. Alex. J. Agric. Res. 51:171-177.

3. Agarwal, V and Chauhan, BM 1988. A study on composition and hypolipidemic effect of dietary fibre from some plant foods. Plant Foods and Human Nutrition 38 (2), 189-197.

4. Ahmad, VU; Arif S; Amber, AR; Usmanghani, K and Miana, GA 1985. A new spermidine alkaloid from Capparis decidua, Heterocycles, 23(12): 3015-3020.

5. Ahmad, VU; Ismail, N; Arif, S and Amber, AVR 1992. Two new N-acetylated spermidine alkaloids from Capparis decidua. J. Nat. Prod. 55: 1509-1512.

6. Ahmed, UV; Ismail, N; Arif, S and Amber AR 1989. Isocodonocarpine from Capparis decidua, Phytochem, 28(9):2493-2495.

7. Ahmed, UV; Ismail, N; Arif, S and Amber, AR 1987. Capparisinine, a new alkaloid from Capparis decidua, Liebigs Ann. Chem, 2:161-162.

8. Ali, S.A., A.A. Gameel, A.H. Mohamed and T. Hassan, 2011. Hepatoprotective activity of Capparis deciduas aqueous and methanolic stems extracts against carbon tetrachloride induced liver histological damage in rats. J. Pharmacol. Toxicol., 6: 62-68.

9. Ali, S.A., T.H. Al-Amin, A.H. Mohamed and A.A. Gameel, 2009. Hepatoprotective activity of aqueous and methanolic extracts of Capparis decidua stems against carbon tetrachloride induced liver damage in rats. J. Pharmaco. Toxicol., 4: 167-172.

10. Anon 2001. Ker: a genetic Wealth of Haryana. CCS HAU, Research Courier 6(1): 4.

11. Anon. 1992. The Wealth of India, Vol.3, Council of Scientific and Industrial Research publication, New Delhi, India: 210-212.

12. Arora, P; Kumar, S; Sharma M K and Mathur S P 2007. Corrosion Inhibition of Aluminium by Capparis decidua in Acidic Media. E-Journal of Chemistry 4 (4): 450-456.

13. Bangarwa, KS 2008. Exploring Capparis decidua for livelihood and wasteland development. Asia-Pacific Agroforestry Newsletter 32:3-5. (Published by FAO/RAP office Philippines).

14. Bhargava, R., N. Vishal and O.P. Pareek. 2000. Note on role of plant growth inhibitor in sprouting of Capparis decidua cuttings. Curr. Agri. 24(1): 131-133.

15. Bhargava, R., P. Verma, P.L. Saroj, and N. Chauhan 2006. Propagation of Capparis decidua. Indian Forester 132(6): 737-745.

16. Chahar, OP; Kharb, P; Ali, SF; Batra, P and Chowdhury, VK 2010 Development of protocol on micropropagation of Ker (Capparis decidua (Forsk) Edgew). World Applied Sciences journal 10(6):695-698.

17. Chahlia, N., 2009. Comparative evaluation of the hypoglycaemic activity of various parts of Capparis decidua. Biharean Biologist., 3: 13-17.

18. Chahlia, Neelkamal 2009. Effect of Capparis decidua on hypolipidemic activity in rats. Journal of Medicinal Plants Research 3(6): 481-484.

19. Chauhan, BM; Duhan, A and Bhat, CM 1986 Nutritional value of Ker (Capparis decidua) fruit, J. of food sci. & technol., India, 23: 106-108.

20. Chouhan, F; Wattoo, M H S; Tirmizi, S; Memon, A F Z; Aziz-Ur-Rahman and Tufail, M 2002 Analytical Investigation of Inorganic Nutritive Elements of Capparis Decidua Grown in Cholistan Desert. The Nucleus, 39 (3-4) 2002: 195-199.

21. Dahot, MU 1993 Chemical evaluation of the nutritive value of flowers and fruits of Capparis decidua, J. Chem. Soc. Pak, 15(1):78-81.

22. Deora, N.S. and N.S. Shekhawat. (1995). Micropropagation of Capparis deciduas (Forsk.) Edgew.: a tree of arid horticulture. Plant cell reports 15(3-4): 278-281.

23. Dhar, DN, Tiwari, RP, Tripathi, RD and Ahuja AP 1972 Chemical examination of Capparis decidua. Proc. Natl. Acad. Sci. Ind. Sect. A, 42:24-27.

24. Duhan, A; Chauhan, BM and Punia D 1992 Nutritional value of some nonconventional plant foods of India. Plant Foods Hum. Nutr., 42: 193-200.

25. Fageria, MS; Khandelwal, P and Dhaka, RS 2003 Effect of harvest stage and processing treatment on quality of sun dried ker (Capparis decidua) fruit. J of Horticultural Science & Biotechnology 78(2):168-172.

26. Gaind, KN; Juneja, TR and Bhandarkar PN 1972. Volatile principle from seeds of Capparis decidua. Kinetics of in vitro antibacterial activity against Vibrio cholerae ogava, inaba, and eltor. Ind. J. Pharm. 34:86-88.

27. Gaind, KN and Juneja TR 1970 Capparis decidua- Phytochemical study of flowers and fruits, Res Bull Punjab Univ Sc, 21:67-71.

28. Gaind, KN; Juneja TR and Jain PC 1969. Anthelmintic and Purgative Activity of Capparis decidua Edgew. Indian J Hosp Pharm; 2: 153-155

29. Gaind, KN; Juneja TRand Jain PC 1969. Investigations on Capparis decidua Edgew. Part II. Antibacterial and antifungal studies. Indian J Pharm; 31: 24-25

30. Ghulam, S 2002 The Phytochemical and Phytopharmacological Studies on Saraca indica, Capparis deciduas and Lotus gracini, Pakistan Research Repository, University of Karachi, Karachi.

31. Govind, K; Vyas R. Sharma, Vinod Kumar; Sharma, T B and Khandelwal, V 2009 Diversity analysis of Capparis decidua (Forssk.) Edgew. using biochemical and molecular parameters. Genet Resour Crop Evol 56:905–911.

32. Goyal, M and Sharma, SK 2009 Traditional wisdom and value addition prospects of arid foods of desert region of North West India. Indian J Traditional Knowledge 8(4):581-585.

33. Goyal, M.; Nagori, B.P and Sasmal, D 2009 Sedative and anticonvulsant effects of an alcoholic extract of Capparis decidua. J. Nat. Med., 63: 375-379.

34. Goyal, R and Grewal RB 1999 Effect of Teent (Capparis decidua) supplementation on levels of blood plasma glucose and total protein hyperlipidemic subject. Haryana J of Horticultural Science 28(3&4:202-204.

35. Goyal, R and Grewal RB 2003 The influence of Teent (C. decidua) on human plasma triglycerides, total lipids and phospholipids, Nutr. Health, 17(1):71-76.

36. Goyal, R and Grewal RB 2005. The influence of teent (Capparis decidua) on human plasma triglycerides, total lipids and phospholipids. Indian J Exp Biol. 3(10):863-6

37. Gupta, I.C; Harsh, LN; Shankaranaryana, KA and Sharma, BD 1989. Wealth from wastelands. Indian Fmg. 38 (11): 18-19.

38. Gupta, J and Ali, M 1998 Phytoconsitituents of Capparis decidua root barks. J Medicinal Aromatic plant sci. 20: 683-689.

39. Gupta, J and, Ali, M 1997 Oxygenated heterocyclic constituents from Capparis decidua root bark, Indian J. Heterocycl Chem, 6(4):295-302.

40. Gupta, M., U. Kanti Mazumder, T. Siva Kumar, P. Gomathi and R. Sambath Kumar, 2004. Antioxidant and hepatoprotective effects of Buhinia racemosa against paracetamol and carbon tetrachloride induced liver damage in rats. Iran J. Pharmacol. Therap., 3: 12-20.

41. Harsh, LN and Tiwari, JC 1998. Biodiversity of vegetational complex in arid regions of India. In: Biodiversity of forest species, R. Bawa and P.K. Khosla (ed), Bishen Singh Mahender Pal Singh, Dehradun, India: 91-104.

42. Iqbal, H; Khan, Z and Zaman, S 2008. Antimicrobial activity of Capparis decidua (Forssk.) Edgew. Pak. J. Plant Sci 14(1): 29-34.

43. Joseph, B and Jini, D 2011. A Medicinal Potency of Capparis Deciduas-A Harsh Terrain Plant. Research j of Phytochemistry 5(1):1-13.

44. Khan, UG 1980. Phytochemical studies on Capparis deciduas and Prosopis glandulosa. PhD Thesis, Univ. Karachi, Karachi, Pakistan.

45. Kumar, NK and Bhandari, MM 1993. Impact of human activities on the pattern and process of sand dune vegetation in the Rajasthan Desert. Desertification Bull. 22: 45-54.

46. Lal, G and Dhaka, RS 2005. Effects of curing treatments on the quality of dried fruits of ker Capparis decidua Linn.). Journal of Food Science and Technology 42(1), 106-108.

47. Mahla, H R; Mertia, RS and Sinha, NK 2010. Morphological characterization of in-situ variability in kair (Capparis decidua) and its management for biodiversity conservation in Thar desert. m Open Access Journal of Medicinal and Aromatic Plants Vol 1, No 2

48. Mali, RG; Hundiwale, JC; Sonawane, RS; Patil, RN and Hatapakki BC 2004. Evaluation of Capparis decidua for anthelmintic and antimicrobial activities, Ind. J. Nat. Prod, 20(4):10-13.

49. Mertia, RS 2001. A new record of yellow flowered Capparis deciduas (Forsk) Edgew from Jaisalmer.Ann. Arid Zone 40(2):219-220.

50. Mishra, SN; Tomar, PC and Lakra, N 2007. Medicinal and food value of Capparis—a harsh terrain plant. Indian J Traditional Knowledge 6(1):230-238.

51. Pandey, AN and Rokad, MV 1992. Sand dune stabilization: an investigation in the Thar desert of India. Journal of arid environments. 22(3): 287-292.

52. Pandey, RP and Shetty, BV 1985. Rare and threatened plants of Rajasthan. In: Proc. Nat.Symp. on Evaluation of Environment Species, S.D. Mishra; D.N.Sen and J.Ahmed (eds). Geobios Univ., Jodhpur, India: 238-241.

53. Panikkar, AON 1961. Chromosome numbers in Capparis decidua. Current Science 31:32.

54. Pareek, OP; Sharma, BD and Sharma, S 1998. Wasteland Horticulture. Malhotra Publishing House, New Delhi, India, 136 pp.

55. Pokharkar Raghunath D; Funde Prasad E and Pingale Shirish S 2007. Aqueous Extract of Capparis Deciduas in Acute Toxicity Effects of the Rat by Use of Toothache Reliever Activity. Pharmacologyonline 3: 511-517.

56. Purohit, A and Vyas, KB 2005. Hypolipidaemic efficacy of Capparis deciduafruit and shoot extract in cholesterol- fed rabbits, Ind. J. Exp. Biol, 43(10):863-866.

57. Purohit, A and Vyas, KB 2006. Antiatherosclerotic effect of Capparis deciduafruit extract in cholesterol- fed rabbits, Pharm Bio, 44(3), 2006, 172-177.

58. Purohit, A and Vyas, KB 2006. Antiatherosclerotic efficacy of Capparis decidua flower extract in cholesterol fed rabbits. Journal of Cell and Tissue Research Vol. 6 (1) 533-536

59. Qaiser M., Qadir SA 1972. A Contribution to the autecology of Capparis decidua. Part 2. Effect of eEdaphic and biotic factors on growth and abundance. Pak. J. Bot. 4: 137-156.

60. Qaiser, M and Qadir, SA 1971. A contribution to the autecology of Capparis decidua (Forssk.) Edgew. Pakistan J. Bot. 3: 37-60.

61. Rai, S and Rai, S 1987. Oils and fats in arid plants with particular reference to Capparis decidua. Indian Soc. Des. Technol. 12:99-105.

62. Rashid, S; Lodhi, F; Ahmad M and Usmanghani, K 1989. Preliminary cardiovascular activity evaluation of capparidisine, a spermidine alkaloid from Capparis decidua. Pak J Pharmacol; 6: 6-16.

63. Rathee, S; Moglal, OP; Sardana, S. Vats, M and Rathee, P. 2010 Antidiabetic activity of Capparis decidua Forsk Edgew. Journal of Pharmacy Research 3(2):231-234.

64. Rathee, Sushila; Mogla, O P; Rathee, Permender and Rathee, Dharmender 2010 Quantification of β- Sitosterol using HPTLC from Capparis decidua (Forsk.) Edgew. Der Pharma Chemica, 2010, 2(4): 86-92.

65. Rathee, Sushila; Rathee, Permender; Rathee, Dharmender; Rathee, Deepti and Kumar, Vikash 2010 Phytochemical and Pharmacological Potential of Kair (Capparis Decidua). International Journal of Phytomedicine 2:10-17

66. Rathore, Mala 2009. Nutrient content of important fruit trees from arid zone of Rajasthan. Journal of Horticulture and Forestry 1(7):103-108.

67. Samadia, DK and Vashishtha, BB 2001. Genetic bio-diversity of ker (Capparis decidua) and lasora (Cordia myxa) in arid and semi arid regions of Rajasthan (Short communication). In: symposium on plant genetic resources management: Advances and challenges held at NBPGR, New Delhi, August 1-3, 2001.

68. Satyanarayana, T; Mathews, AA and Vijetha, P 2008. Phytochemical and pharmacological review of some Indian Capparis Species. Pharma. Rev., 2: 36-45.

69. Shankarnaryan, KA; Harsh, LN and Kathju, S 1987. Agroforestry in the arid zones of India. Agroforestry Systems. 5: 69-88.

70. Sharma, B and Kumar, P 2009. Extraction and Pharmacological Evaluation of Some Extracts of Tridax procumbens and Capparis deciduas. International Journal of Applied Research in Natural Products 1(4):5-12.

71. Sharma, A; Sharma M S; Mishra, A; Sharma, S; Kumar, B and Bhandari, A 2011. A REVIEW ON THAR PLANTS USED IN LIVER DISEASES. INTERNATIONAL JOURNAL OF RESEARCH IN PHARMACY AND CHEMISTRY 1(2):224-236.

72. Sharma, B; Salunke, R; Balomajumder, C; Daniel, S and Roy, P 2010. Anti-diabetic potential of alkaloid rich fraction from Capparis decidua on diabetic mice. Journal of Ethnopharmacology 127 :457–462.

73. Sharma, I; Gusain, D; Sharma, A and Dixit, VP 1991. Hypolipidaemic effect of Capparis decidua fruit extract (50% EtOH) in cholesterol-fed rabbits. Indian Drugs; 28: 412-416.

74. Shekhawat, JS 1994. Studies on growth, fruit development and quality of ker (Capparis decidua (Forsk) Edgew). Ph.D. thesis, Rajasthan Agricultural University, Bikaner, Rajasthan, India.

75. Shikha, B; Prakash, NB and Nikhil, B 2011. Medicinal Plants of Rajasthan (India) with. Antidiabetic Potential. International Res. J Pharmacy 2(3):1-7.

76. Singh, D and Singh, RK 2011. Kair (Capparis decidua): A potential ethno botanical weather predictor and livelihood security shrub of arid zone of Rajasthan and Gujrat. Indian J Traditional Knowledge 10(1):146-155.

77. Tyagi, P and Kothari, SL 2001.Continuous shoot production for micropropagation of Capparis decidua - a tree of arid agroforestry system. - J. indian bot. Soc. 80: 5-8.

78. Tyagi, P and Kothari, SL1997. Micropropagation of Capparis deciduas through in vitro shoot proliferation on nodal explants of mature tree and seedling explants. J. Plant Biol. Biotechnol. 6: 19-23.

79. Tyagi, P; Khanduja, S and Kothari, SL 2005. Somatic embryogenesis in Capparis deciduas (Forsk) Edgew. - a multipurpose agroforestry plant. - J. Plant Biochem. Biotechnol. 14: 197-200.

80. Tyagi, P; Khanduja, S and Kothari, SL 2010. In vitro culture of Capparis decidua and assessment of clonal fidelity of the regenerated plants. Biologia Plantarum 54 (1): 126-130.

81. Upadhyay RK, Rohatgi L, Chaubey MK, Jain SC 2006. Ovipositional responses of the pulse beetle, Bruchus chinensis (Coleoptera: Bruchidae) to extracts and compounds of Capparis decidua. J. Agric. Food Chem. 54: 9747-9751.

82. Upadhyay, R K Shoeb Ahmad, Rajani Tripathi, Leena Rohtagi and Subhash C. Jain 2010. Screening of antimicrobial potential of extracts and pure compounds isolated from Capparis deciduas. Journal of Medicinal Plants Research Vol. 4(6):439-445.

83. Upadhyay, R K; Jaiswal, G and Ahmad, S 2010 Anti-termite efficacy of Capparis decidua and its combinatorial mixtures for the control of Indian white termite Odontotermes obesus (Isoptera: Odontotermitidae) in Indian soil. J. Appl. Sci. Environ. Manage. 14 (3) 101 – 105.

84. Vyas, GK; Sharma, R; Kumar, V; Sharma, TB and Khandelwal, V 2009. Diversity analysis of Capparis decidua (Forsk.) Edgew. using biochemical and molecular parameters. Genetic Resources and Crop Evolution 56(7):905-911.

85. Yadav, P., Sarkar, S., Bhatnagar, D. 1997a. Action of Capparis decidua against alloxan-induced oxidative stress and diabetes in rat tissues. Pharmacological Research 36 (3), 221_ 228.

86. Yadav, P., Sarkar, S., Bhatnagar, D. 1997b. Lipid peroxidation and antioxidant enzymes in erythrocytes and tissues in aged diabetic rats. Indian Journal of Experimental Biology 35 (4), 389-392.

Executive Summary of the Project

Capparis Linn. (Capparidaceae) is the major genus of climbing shrubs, bushes or small trees and 26 species of this genus are reported to occur in India (Anon., 1992). *Capparis decidua* (Forsk.) Edgew. Syn. *C aphylla* Roth commonly known as kair, is an important medicinal plant. It is a densely branched, spinous shrub or tree, upto 6 m in height (rarely 10 m), with a clear bole of 2.4 m, found chiefly in the dry, arid and semi arid regions of India extending from Punjab southwards to Tamil Nadu. This species is common in dry tropical Africa, especially in the Sahel, where it sometimes constitutes lines of small trees in Wadi beds, as in Mauritania for instance. In West Africa, the area of distribution is identical to that of Cadaba farinosa; its southern limit corresponds to the northern loop of the Senegal river. In the Republic of Niger, it reaches the Konadougou. Its area includes Tibesti (West Chad), much of the Sudan (except the extreme South) the Arabian Peninsula, Jordan, India, Pakistan, Iran, the Mascarene Islands and Natal. It is tolerant to prolonged drought and an interesting plant by reason of its excellent adaptation to arid conditions. It is found in varied habitats and has very good soil binding capacity which can be propagated and cultivated on large scale for checking wind erosion on sandy wastelands (Gupta *et al.*, 1989). Capparis decidua can be used in landscape gardening, afforestation and reforestation in semi desert and desert areas; it provides assistance against soil erosion (Joseph and Jini, 2011). It is second to none as for as its capability to stabilize sands and combat drought with fair tolerance to salinity and alkalinity is concerned. It also improves the fertility of sand dunes and reduces alkalinity very sharply. In tropical and sub-tropical regions plant occurs within annual rainfall 100 to 400 mm and temperature 16° to 50°C.

Capparis decidua is a shrub and an important medicinal plant. It is naturally found in varied habitats and has very good soil binding capacity. It improves the fertility of sand dunes and reduces alkalinity very sharply. Pickle and cooked vegetable of its unripe fruits form an integral part of human diet for stomach troubles in tropical and sub-tropical regions. Unripe green fruits are sold as hot cake in different regions. Top shoots and young leaves of the plant are used as plaster for boils and swelling, in powder form to relieve tooth troubles including pyorrhea. Fruits and seeds are used in cholera, dysentery and urinary purulent discharges. Root parts are useful in boils, eruptions, swelling, chronic and foul ulcers, cough and asthma.

Wood is hard, heavy, and resistant to termites. Due to the xerophytic nature and usefulness in sustenance for humanity, Capparis decidua is suggested for the preservation (Joseph and Jini, 2011). Capparis decidua checks the sand dune formation and soil erosion in addition to medicinal and food logistics. The tendency of high networking of root system of Capparis decidua apparently facilitates the weathering of stone/rocks and soil formation. Therefore, there is a need to exploit plant species for afforestation in the hot dry to cold deserts regions.

The project work on **"Plant Diversity Assessment and Establishment of Germplasm Bank for Conservation, Domestication and Breeding in *Capparis decidua*"** under Dr B.P. Pal National Environment Fellowship Award for Biodiversity was started in December 2008 for two years with following objectives:

i. Assessment of plant diversity, rate of loss in plant diversity and causes for decline in biodiversity of *Capparis decidua*.

ii. Documentation of traditional knowledge, folk varieties and other notable material.

iii. Identification of promising reproductive material of *Capparis decidua* from different fragile environmental conditions.

iv. Establishment of germplasm bank of *Capparis decidua*.

v. Domestication and breeding of *Capparis decidua*.

The results of the project work are summarized below:

- A survey of southern Haryana and Rajasthan was conducted during flowering and fruiting seasons. *Capparis decidua* has its existence in undisturbed areas. *Salvadora oleoides* has been very closely associated tree species of *Capparis decidua*. *Prosopis cineraria, Azadirachta indica* and *Calotropis procera* were also observed to be associated species. *Prosopis juliflora* an exotic tree species having very fast rate of growth is spreading like weeds particularly in undisturbed lands. In this way, undisturbed lands are common habitat for *Capparis decidua* and Prosopis *juliflora*. *Prosopis juliflora* (which regenerates faster and grows aggressively) has become a serious threat for the existence of our indigenous plants/trees and *Capparis decidua* is a serious victim of *Prosopis juliflora*. The flower buds and unripe green fruits of *Capparis decidua* are pickled and also cooked and eaten as vegetable. The immature fruits are in great demand and have high economic value. Hence, the immature fruits are harvested and sold at high prices (upto Rs sixty per Kg). The local people used to go for over exploitation of *Capparis decidua*. They usually try to collect fruits from inner depth of shrub even by damaging the branches. The availability of matured seed is almost nil. This practice puts seed production and propagation of *Capparis decidua* at risk. Poor or lack of seed production continues to be major cause for declining population of *Capparis decidua*. Therefore, regeneration in *Capparis decidua* is through suckers only and that too is suppressed by *Prosopis juliflora*. Hence, *Prosopis juliflora* is seriously threatening the survival of *Capparis decidua* and other indigenous species.

- Flowering and fruiting in *Capparis decidua* have been observed twice in a year. Flowering was observed during March-April and by the end of April fruit setting was completed. But in some parts, flowering was observed up to end of May. Regular flowering was also observed during August-September and by the end of September fruit setting was completed. Lot of variation was observed for flowering and fruiting between regions and between plants with in region. From a group of *Capparis decidua* plants/shrubs/trees at a particular place, some plants were observed to have full flowering whereas other plants were observed to have flowering intensity ranging from zero to 80 %. The flowers of *Capparis decidua* were observed to be complete having all the four parts viz., calyx (four sepals), corolla (four petals), androecium (many stamens usually 13-16 with a length varies from 0.5 to 1.2 cm) and gynoecium (ovary superior, bicarpellary syncarpous, unilocular with many ovules arranged on parental placenta, style absent and stigma about 1.1 cm.). In *Capparis decidua* stigma is closely surrounded by anthers. Pollination may be before flower opening or after flower opening. But the position of anthers in relation to stigma seems to ensure self-pollination. Sufficient no of fruits were observed in a *Capparis decidua* plants in isolation. Hence, *Capparis decidua is* predominantly self-pollinating species. The seeds collected from isolated plants of *Capparis decidua* showed normal growth. These observations further confirmed the self-pollinating nature of *Capparis decidua*.

- The seeds of *Capparis decidua* showed lot of variation for seed germination (fresh seeds) ranging from zero percent to 63.89%. The observations clearly suggested the extent of variation for seed dormancy among different seed lots of *Capparis decidua*. More than 50 % seed germination was observed upto one year.

- Index score analysis on the basis of simultaneous consideration of medicinal value, food value and potential for sand dune stabilization and salt tolerance suggested the superiority of *Capparis decidua* (0.91) in comparison to *Azadirachta indica* (0.86), *Prosopis cineraria* (0.85) and *Acacia nilotica* (0.84). *Azadirachta indica* was found slightly better than Capparis decidua on the basis of medicinal importance. *Prosopis cineraria* was observed to be slightly better than *Capparis decidua* on the basis of food value. *Acacia nilotica* and *Capparis decidua* have equal potential for salt tolerance whereas *Capparis decidua* and *Prosopis cineraria* have equal potential for sandy region. Index score analysis on the basis of simultaneous consideration of medicinal value, food value and potential for sand dune stabilization and salt tolerance suggested the superiority of *Capparis decidua* in comparison to *Azadirachta indica*, *Prosopis cineraria* and *Acacia nilotica*. *Capparis decidua* and *Prosopis cineraria* have almost equal hardiness to bear extreme conditions of arid region. Moreover, *Capparis decidua* can be successfully grown readily from seed and root suckers. Seed production is a serious problem as both *Capparis decidua* and *Prosopis cineraria* are over exploited because of highly economic importance of their immature fruits.

- The flower buds and unripe green fruits of *Capparis decidua* are pickled and also cooked and eaten as vegetable. Pickle and cooked vegetables of unripe fruits are very useful for stomach troubles especially for constipation.

- The bark of *Capparis decidua* has been shown to be useful in the treatment of coughs, asthma, bronchial inflammation, indigestion and rheumatism. The stem bark decoction (10-15ml) is administered twice a day in asthma and other respiratory disorders. The stem bark is used as a laxative, diaphoretic and anthelmintic. The bark of its leafless shrub is used for the treatment of asthma, cough, inflammation and acute pain. The stem is used in pyorrhea and rheumatism. Powered coal of stem with water is taken for the treatment of fractured bone. The bark has an acrid, sharp, hot taste; analgesic, diaphoretic, alexeteric, laxative, in dropsy ground, anthelmintic; good in asthma, ulcers and boils, vomiting, piles and all inflammations.

- The root and root bark are pungent and bitter and are given to treat intermittent fevers and rheumatism. It is applied externally to ribs in case of pleurisy. The root bark extract is given twice a day for 3 days in the treatment of haemorrhoids. Charcoal of root is taken orally for rheumatism and bone fracture. The inner bark of roots is used to treat scabies and eczema. Root paste is applied on scorpion bite.

- Paste of coal from wood is applied extremely to muscular injuries. Juice of fresh plant is dropped into the ear to kill worms. The tender branches and leaves are used as a plaster for boils and swellings and to relieve toothache on chewing. Paste made of aerial part is applied on fracture as an analgesic and anti-inflammatory. The top shoots and young leaves are made into a powder and used as a blister; they are also used in boils, eruptions and swellings and as an antidote to poison.

- The extract of unripe fruits and shoots are useful for controlling heart, liver and kidney problem. The fruits of *C. decidua* have medicinal value and are believed to provide relief from cardiac and gastric troubles. Many vaids prescribe ker fruits for cardiac trouble. The extract of immature fruits can be used to cure trachoma (Chronic conjunctivitis that can cause blindness). The powered fruit of *Capparis decidua* is used in anti-diabetic formulations. The fruit has a sharp hot astringent to the bowels; destroys foul breath, biliousness, and urinary purulent discharges; good in cardiac troubles. Floral primordial of *Capparis decidua,* boiled in oil and tied between the infected hooves, is found to be curative

- The *Capparis decidua* has been traditionally useful in the treatment of coughs, asthma, bronchial inflammation, respiratory disorders, indigestion, pyorrhea, intermittent fevers, rheumatism, ulcers, boils, vomiting, piles, bone fracture inflammation and acute pain, scorpion bite, antidote to poison, toothache, diabetes, heart, liver and kidney problems, trachoma (Chronic conjunctivitis that can cause blindness), infected hooves, scabies and eczema. It is being used as laxative, tonic and mouth freshner.

- *Capparis decidua* is of much use in climate prediction and features in farmers' strategies in natural resources production and management and agricultural planning. The local farmers of Surender Nagar, Kheda, Bhavnagar and Ahmedabad of Gujarat state that they use kair (locally called kerda) in predicting the weather, namely temperature and rainfall. Apart from the immense use of kerda in making local pickle, the farmers consider that if blooming in kerda is greater and flowers are deep pink, then the temperature is more than 45°C and the rainfall will be less than normal. Based on the observations of the numbers of flowers and fruits and the canopy of kerda, the farmers select their crop varieties and cropping systems for the following rainy season.

- Large sized trees up to 8.5 metres with higher yield of fruit per plant were observed in Salasar, Sikar (Rajasthan), Kalwas, Hisar (Haryana) and Balsamand, Hisar (Haryana). Plants mostly shrubs with large spread and higher yield of fruit per plant were obsereved in Sahewala, Hisar (Haryana), Bapoda, Bhiwani (Haryana), Fatehpur, Sikar (Rajasthan), Salasar, Sikar (Rajasthan), Ghodaria Khurad, Sikar (Rajasthan), Nokhda, Bikaner (Rajasthan), Phagalwas, Sikar (Rajasthan) and Dadar, Churu (Rajasthan).

- Seedlings of nine progenies were transplanted in the field following Randomized Block Design (RBD) with three replications during September, 2009. The plant height varies from 31.2 cm to 43.5 cm with a mean of 37.5 cm whereas plant spread varies from 27.5 cm to 39.4 cm with a mean of 35.8 cm at the age of fifteen months after transplanting. The root length was observed higher than plant height and plant spread. The root length varies from 39.4 cm to 86.5 cm with a mean of 65 cm. Root length varies from 1.26 times of plant height to 2.04 times of plant height with a mean of 1.73 times of plant height whereas root length varies from 1.43 times of plant spread to 2.06 times of plant spread with a mean of 1.81 times of plant spread. Coefficient of variation for root length was 25.7 percent whereas for plant height and plant spread, coefficients of variations were 11.06 per cent and 15.7 per cent, respectively.

- Seedlings of thirty-nine accessions collected from southern Haryana, adjoining Rajasthan and Bikaner region of Rajasthan were transplanted in the field following RBD during February, 2010. The highest plant height of 24 cm was observed in progeny no 51 from Kalwas, Hisar (Haryana) followed by progeny no 17 and 16 from Nokhda, Bikaner (Rajasthan), progeny no 40 from Bapoda, Bhiwani (Haryana), progeny no 28 from Sahewala, Hisar (Haryana) and progeny no 4 from Hawala, Rajsamand (Rajasthan) nine months after transplanting. The progeny no 51 from Kalwas, Hisar (Haryana) and progeny no 28 from Sahewala, Hisar (Haryana) were observed comparatively more erect growing.

- Seedlings of seventy-seven accessions collected from southern Haryana, adjoining Rajasthan and Bikaner region of Rajasthan were transplanted in the field following Randomized Block Design (RBD) with three replications during September, 2010. The highest plant height of 20.6 cm was observed in accession no 142 from Salasar, Sikar (Rajasthan) followed by accession no 51 from Kalwas, Hisar (Haryana), accession no 40 from Bapoda, Bhiwani (Haryana), accession no 16 from Nokhda, Bikaner (Rajasthan), accession no 118 from Nokhda, Bikaner (Rajasthan) and accession no 128 from Fatehpur, Sikar (Rajasthan) three months after transplanting. The plant height at the age of three months after transplanting varies from 11.4 cm in accession no 33 from Loharu, Bhiwani (Haryana) to 20.6 cm in accession no 142 from Salasar, Sikar (Rajasthan) with a mean of 15.52 cm and 13.55 % coefficient of variation.

- *Capparis decidua* can be successfully grown from seeds and suckers. Its matured fruits which have turned pink should be collected either in May-June or August-October. Seeds should be separated out by washing the pulp of fruits with in 2-3 days. This will enhance the viability of seeds. Seeds are to be sown in polythene bags (with 3:1 ratio of sandy soil and FYM) in August if seeds are collected in May-June or February if seeds are collected in August-October. Collection of seeds in August-October and sowing of seeds in early February are desirable. Transplanting of seedlings should be done in early February by choosing seedlings of more than 15 cm height.

- Suckers of *Capparis decidua* are to be collected in early February and should be established in polythene bags (with 3:1 ratio of sandy soil and FYM). Plants raised from suckers attain usually attain a height of more than 15 cm in 4-5 months and hence these plants are transplanted in field during rainy season.

(Part-III)

Recommendation including remedial measures relevant to the environmental problems studied under the scheme

- *Capparis decidua* has proven to be an economically important plant in southern Haryana, Rajasthan and elsewhere. It provides varied food and medicinal uses, building materials, fuel wood, and other income-generating opportunities. It contributes to environmental sustainability due to its soil-binding capacity and its ability to improve the soil fertility of sand dunes and to reduce soil alkalinity. For large scale plantation of *Capparis decidua,* it has been established that *Capparis decidua* can be successfully grown from seeds and suckers. Its matured fruits which have turned pink should be collected either in May-June or August-October. Seeds should be separated out by washing the pulp of fruits within 2-3 days. Seeds are to be sown in polythene bags (with 3:1 ratio of sandy soil and FYM) in August if seeds are collected in May-June or February if seeds are collected in August-October. Collection of seeds in August-October and sowing of seeds in early February are desirable. Transplanting of seedlings should be done in early February by choosing seedlings of more than 15 cm height. Suckers of *Capparis decidua* are to be collected in early February and should be established in polythene bags (with 3:1 ratio of sandy soil and FYM). Plants raised from suckers attain usually attain a height of more than 15 cm in 4-5 months and hence these plants are transplanted in field during rainy season.

- As per the survey of traditionally usefulness, it is recommended that *Capparis decidua* may be used for treatment of coughs, asthma, bronchial inflammation, respiratory disorders, indigestion, pyorrhea, intermittent fevers, rheumatism, ulcers, boils, vomiting, piles, bone fracture inflammation and acute pain, scorpion bite, antidote to poison, toothache, diabetes, heart, liver and kidney problems, trachoma (Chronic conjunctivitis that can cause blindness), infected hooves, scabies and eczema. It may be used as laxative, tonic and mouth freshner.

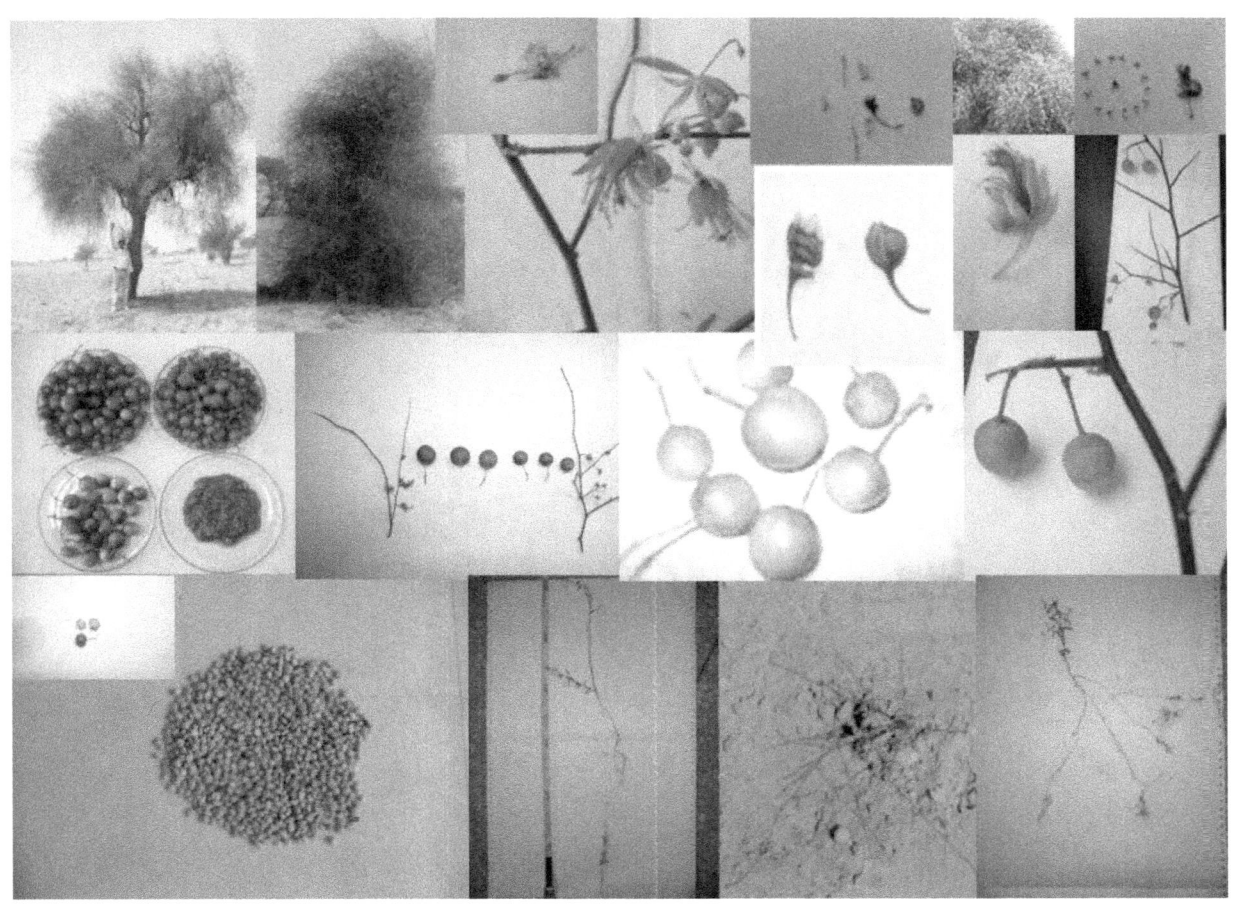

www.ingramcontent.com/pod-product-compliance
Lightning Source LLC
Chambersburg PA
CBHW082112220526

45472CB00009B/2154